VOLUME ONE HUNDRED AND THIRTY TWO

ADVANCES IN COMPUTERS

Applying Computational Intelligence for Social Good

VOLUME ONE HUNDRED AND THIRTY TWO

ADVANCES IN COMPUTERS

Applying Computational Intelligence for Social Good

Edited by

PREETHA EVANGELINE DAVID
Department of Artificial Intelligence and Machine Learning,
Chennai Institute of Technology,
Chennai, Tamil Nadu, India

P. ANANDHAKUMAR
Department of Information Technology,
Madras Institute of Technology, Anna University,
Chennai, Tamil Nadu, India

Academic Press is an imprint of Elsevier
125 London Wall, London, EC2Y 5AS, United Kingdom
525 B Street, Suite 1650, San Diego, CA 92101, United States
50 Hampshire Street, 5th Floor, Cambridge, MA 02139, United States

First edition 2024

ISBN: 978-0-323-88544-7
ISSN: 0065-2458

For information on all Academic Press publications
visit our website at https://www.elsevier.com/books-and-journals

Publisher: Zoe Kruze
Editorial Project Manager: Palash Sharma
Production Project Manager: James Selvam
Cover Designer: Greg Harris

Typeset by STRAIVE, India

Contents

Contributors

K.A. Alex Luke
Department of Information Technology, Madras Institute of Technology, Anna University, Chennai, Tamil Nadu, India

P. Anandhakumar
Department of Information Technology, Madras Institute of Technology, Anna University, Chennai, Tamil Nadu, India

Gayathri Ananthakrishnan
Department of Information Technology, VIT University, Vellore, India

Erik Cambria
School of Computer Science and Engineering, Nanyang Technological University, Singapore, Singapore

V. Cephas Paul Edward
Anna University, Chennai, Tamil Nadu, India

Pethuru Raj Chelliah
Reliance Jio Platforms Ltd., Bangalore, India

Preetha Evangeline David
Department of Artificial Intelligence and Machine Learning, Chennai Institute of Technology, Chennai, Tamil Nadu, India

E. Gokkul
Department of Computer Technology, Madras Institute of Technology (MIT) Campus, Anna University, Chrompet, Chennai, India

Ashok Kumar Jayaraman
Department of Information Science and Technology, Anna University, Chennai, India

Lakshminarayana Kodavali
Department of CSE, Koneru Lakshmaiah Education Foundation (KLEF), Vadeswaram, Guntur, India

S. Malathi
Department of Artificial Intelligence and Data Science, Panimalar Engineering College, Chennai, India

S. Muthurajkumar
Department of Computer Technology, Madras Institute of Technology (MIT) Campus, Anna University, Chrompet, Chennai, India

B. Rahul
Department of Computer Technology, Madras Institute of Technology (MIT) Campus, Anna University, Chrompet, Chennai, India

R. Ramaprabha
Department of EEE, SSN College of Engineering, Chennai, India

G. Ramya
Department of Computing Technologies, Faculty of Engineering and Technology, SRM Institute of Science and Technology, Kattankulathur, Chennai, India

Atma Sahu
Coppin State University, Baltimore, MD, United States

L.S. Sanjay Kumar
Department of Computer Technology, Madras Institute of Technology (MIT) Campus, Anna University, Chrompet, Chennai, India

K. Sathiyamurthy
Department of CSE, Puducherry Technological University, Puducherry, India

Tina Esther Trueman
Department of Computer Science, University of the People, Pasadena, CA, United States

V. Vivek
School of Computer Science & Engineering, Faculty of Engineering & Technology, JAIN (Deemed-to-be University), Bangalore, India

Preface

Technology has helped societies all over the world to battle the most pressing issues and solve social problems. By promising faster technological advancement, Computational Intelligence promises to provide answers to questions relating to the environment, security, and society that we are all exploring today. There's no denying that CI is on its way to change the world as we know it. There are various social domains such as Environment, Education, Information Security, Healthcare, Crisis Response, Human Behavior and Bias, Disaster Management, and Industrial management, and most recently, Epidemics and Outbreaks are highly in need of Intelligent Technologies to address their issues. This book presents the views on how Computational Intelligent and ICT technologies can be applied to ease or solve social problems by sharing examples of research results from studies of social anxiety, environmental issues, mobility of the disabled, and problems in social safety. As with most changes in life, there will be positive and negative impacts on society as artificial intelligence continues to transform the world we live in. With the idea of future scope in research, the book focuses on the challenges that might be faced (and we should be thinking about how to address them now) as well as several of the positive impacts artificial intelligence will have on society.

The motivation behind writing this book is to bring together the researchers who are keen on working toward socially relevant transdisciplinary research. And today, Computational Intelligence serves as the major scope for researchers, and this book focuses on bringing out the challenges in various domains that could be solved using Intelligent Techniques.

The transformative impact of artificial intelligence on our society will have far-reaching economic, legal, political, and regulatory implications that need discussions and preparations. As more and more data are collected about every single minute of every person's day, our privacy gets compromised, which could devolve into social oppression. The underlying mission is to better understand social interaction and to build machines that work more collaboratively and effectively with humans. When CI takes over repetitive or dangerous tasks, it frees up the human workforce to do work they are better equipped for—tasks that involve creativity and empathy, among others. With better monitoring and diagnostic capabilities, intelligence can dramatically influence healthcare. Our society will gain countless hours of productivity with just the introduction of autonomous transportation and CI, influencing our traffic congestion issues. The way we uncover criminal activity and solve crimes will be enhanced with artificial intelligence. Facial recognition technology is becoming just as common as fingerprints. The use of AI in the justice system also presents many opportunities to figure out how to effectively use the technology without compromising an individual's privacy. At the top of this discussion, it is made clear that we have no predictions about what the future will hold, but we do believe that it is our responsibility to work toward a future we believe in. It is in our power to bring about a future with CI.

Why is implementing computational intelligence for social good so challenging? Principles and its application

Preetha Evangeline David[a] and P. Anandhakumar[b]
[a]Department of Artificial Intelligence and Machine Learning, Chennai Institute of Technology, Chennai, Tamil Nadu, India
[b]Department of Information Technology, Madras Institute of Technology, Anna University, Chennai, Tamil Nadu, India

Contents

Advances in Computers, Volume 132
ISSN 0065-2458
https://doi.org/10.1016/bs.adcom.2023.08.001

Abstract

Computational intelligence (CI) has the potential to help tackle some of the world's most challenging social problems. Real-life examples of AI are already being applied in about one-third of these use cases They range from diagnosing cancer to helping blind people navigate their surroundings, identifying victims of online sexual exploitation, and aiding disaster-relief efforts etc. AI is only part of a much broader tool kit of measures that can be used to tackle societal issues, however. For now, issues such as data accessibility and shortages of AI talent constrain its application for social good. This chapter has grouped use cases into 10 social-impact domains based on taxonomies in use among social-sector organizations. Each use case highlights a type of meaningful problem that can be solved by one or more AI capability. The cost of human suffering, and the value of alleviating it, are impossible to gauge and compare. Nonetheless, employing usage frequency as a proxy, we measure the potential impact of different AI capabilities.

1. Introduction

Computational Intelligence is a new concept for advanced information processing. The objective of Computational Intelligence approaches is to realize a new approach for analyzing and create flexible information processing of humans such as sensing, understanding, learning, recognizing, and thinking.

It is the theory, design, application, and development of biologically and linguistically motivated computational paradigms. Traditionally the three main pillars of CI have been:

- Neural Networks: Using the human brain as a source of inspiration, artificial neural networks (NNs) are massively parallel distributed networks that have the ability to learn and generalize from examples.
- Fuzzy Systems: Using the human language as a source of inspiration, fuzzy systems (FS) model linguistic imprecision and solve uncertain problems based on a generalization of traditional logic, which enables us to perform approximate reasoning.
- Evolutionary Computation: Using biological evolution as a source of inspiration, evolutionary computation (EC) solves optimization problems by generating, evaluating and modifying a population of possible solutions.

However, in time many nature-inspired computing paradigms have evolved. Thus CI is an evolving field and at present in addition to the three main constituents, it encompasses computing paradigms like ambient intelligence, artificial life, cultural learning, artificial endocrine networks, social reasoning, and artificial hormone networks.

2. What is computational intelligence?

CI is the study of intelligent behavior and how to make machines do things at which humans are doing better. CI is one of the technological breakthroughs of this digital era which aims at emulating human intelligence on machines to make them think and behave like human beings. CI is based on the idea that human intelligence could be replicated in computer programs. Although, the idea of creating intelligent machines—ones that are as smart as or smarter than human beings—is not new it became part of modern science with the rise of digital computers and the proliferation of the internet. On the practical side, CI means to create computer programs that perform tasks as well as or better than humans. In simple terms, CI is human intelligence demonstrated by machines.

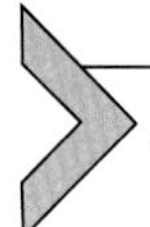

3. What is the difference between artificial intelligence and computational intelligence?

Artificial intelligence (AI) is the study of intelligent behavior demonstrated by machines as opposed to the natural intelligence in human beings. It is an area of computer science that is concerned with the development of a technology that enables a machine or computer to think, behave, or act in a more humane way. Computational Intelligence (CI), on the other hand, is more like a sub-branch of CI that emphasizes on the design, application and development of linguistically motivated computational models. It is the study of adaptive mechanisms to enable or facilitate intelligent behavior in complex and changing environments.

4. Science and engineering

As CI is a science, its literature should manifest the scientific method, especially the creation and testing of refutable theories. Obvious questions are, "What are CI theories about?" and "How would I test one if I had one?" CI theories are about how interesting problems can be represented and solved by machine. Theories are supported empirically by constructing implementations, part of whose quality is judged by traditional computer science principles. You can't accomplish CI without specifying theories and building implementations; they are inextricably connected. Of course, not every researcher needs to do both, but both must be done. An experiment means nothing without a theory against which to evaluate it, and a theory without potentially confirming or refuting evidence is of little use. Ockham's Razor is our guide: Always prefer simple theories and implementations over the more complex. With these thoughts in mind, you can quickly consider one of the most often considered questions that arises in the context of CI: "Is human behaviour algorithmic?" You can dispense with this question and get on with your task by acknowledging that the answer to this question is unknown; it is part of cognitive science and CI's goal to find out.

5. Relationship to other disciplines

CI is a very young discipline. Other disciplines as diverse as philosophy, neurobiology, evolutionary biology, psychology, economics, political science, sociology, anthropology, control engineering, and many more have been studying intelligence much longer. We first discuss the relationship with philosophy, psychology, and other disciplines which study intelligence; then we discuss the relationship with computer science, which studies how to compute. The science of CI could be described as "synthetic psychology," "experimental philosophy," or "computational epistemology"—Epistemology is the study of knowledge. It can be seen as a way to study the old problem of the nature of knowledge and intelligence, but with a more powerful experimental tool than was previously available. Instead of being able to observe only the external behavior of intelligent systems, as philosophy, psychology, economics, and sociology have traditionally been able to do, we are able to experiment with executable models of intelligent behavior. Most importantly, such models are open to inspection, redesign, and experiment in a complete and rigorous way. In other words, you now have a way to construct the models that philosophers could only theorize about. You can experiment with these models, as opposed to just discussing their abstract properties.

Our theories can be empirically grounded in implementation. Just as the goal of aerodynamics isn't to synthesize birds, but to understand the phenomenon of flying by building flying machines, CI's ultimate goal isn't necessarily the full-scale simulation of human intelligence. The notion of psychological validity separates CI work into two categories: that which is concerned with mimicking human intelligence—often called cognitive modeling—and that which isn't. To emphasize the development of CI as a science of intelligence, we are concerned, in this book at least, not with psychological validity but with the more practical desire to create programs that solve real problems. Sometimes it will be important to have the computer to reason through a problem in a human-like fashion. This is especially important when a human requires an explanation of how the computer generated an answer.

Some aspects of human cognition you usually do not want to duplicate, such as the human's poor arithmetic skills and propensity for error. Computational intelligence is intimately linked with the discipline of computer science. While there are many non-computer scientists who are researching CI, much, if not most, CI (or AI) research is done within computer science departments. We believe this is appropriate, as the study of computation is central to CI.

It is essential to understand algorithms, data structures, and combinatorial complexity in order to build intelligent machines. It is also surprising how much of computer science started as a spin off from AI, from timesharing to computer algebra systems. There are other fields whose goal is to build machines that act intelligently. Two of these fields are control engineering and operations research. These start from different points than CI, namely in the use of continuous mathematics. As building real agents involves both continuous control and CI-type reasoning, these disciplines should be seen as symbiotic with CI. A student of either discipline should understand the other. Moreover, the distinction between them is becoming less clear with many new theories combining different areas. Unfortunately, there is too much material for this book to cover control engineering and operations research, even though many of the results, such as in search, have been studied in both the operations research and CI areas.

Finally, CI can be seen under the umbrella of cognitive science. Cognitive science links various disciplines that study cognition and reasoning, from psychology to linguistics to anthropology to neuroscience. CI distinguishes itself within cognitive science because it provides tools to build intelligence rather than just studying the external behavior of intelligent agents or dissecting the inner workings of intelligent systems.

6. Representation and reasoning

In order to use knowledge and reason with it, you need what we call a representation and reasoning system (RRS). A representation and reasoning system is composed of a language to communicate with a computer, a way to assign meaning to the language, and procedures to compute answers given input in the language. Intuitively, an RRS lets you tell the computer something in a language where you have some meaning associated with the sentences in the language, you can ask the computer questions, and the computer will produce answers that you can interpret according to the meaning associated with the language. At one extreme, the language could be a low-level programming language such as Fortran, C++, or Lisp. In these languages the meaning of the sentences, the programs, is purely in terms of the steps the computer will carry out to execute the program. How computation will be carried out given a program and some input, is straightforward to determine. How to map from an informal statement of a problem to a representation of the problem in these RRSs, programming, is a difficult task. At the other extreme, the language could be a natural language, such as English, where the sentences can refer to the problem domain. In this case, the mapping from a problem to a representation is not very difficult: You need to describe the problem in English. However, what computation needs to be carried out in the computer in response to the input is much more difficult to determine. In between these two extremes are the RRSs that we consider in this book. We want RRSs where the distance from a natural specification of the problem to the representation of the problem is not very far. We also want RRSs where the appropriate computation, given some input, can be effectively determined. We consider languages for the specification of problems, the meaning associated with such languages, and what computation is appropriate given input in the languages. One simple example of a representation and reasoning system between these two extremes is a database system. In a database system, you can tell the computer facts about a domain and then ask queries to retrieve these facts. What makes a database system into a representation and reasoning system is the notion of semantics. Semantics allows us to debate the truth of information in a knowledge base and makes such information knowledge rather than just data. In most of the RRSs we are interested in, the form of the information is more flexible and the procedures for answering queries are more sophisticated than in a database. A database typically has table lookup; you can ask about what is in the database, not about what else must be true, or is likely to be true, about the domain.

7. Ontology and conceptualization

An important and fundamental prerequisite to using an RRS is to decide how a task domain is to be described. This requires us to decide what kinds of things the domain consists of, and how they are to be related in order to express task domain problems. A major impediment to a general theory of CI is that there is no comprehensive theory of how to appropriately conceive and express task domains. Most of what we know about this is based on experience in developing and refining representations for particular problems. Despite this fundamental problem, we recognize the need for the following commitments. The world can be described in terms of individuals (things) and relationships among individuals. An ontology is a commitment to what exists in any particular task domain. This notion of relationship is meant to include propositions that are true or false independently of any individuals, properties of single individuals, as well as relationships between pairs or more individuals. This assumption that the world can be described in terms of things is the same that is made in logic and natural language. This isn't a strong assumption, as individuals can be anything nameable, whether concrete or abstract. For example, people, colors, emotions, numbers, and times can all be considered as individuals. What is a "thing" is a property of an observer as much as it is a property of the world. Different observers, or even the same observer with different goals, may divide up the world in different ways. For each task or domain, you need to identify specific individuals and relations that can be used to express what is true about the world under consideration. How you do so can profoundly affect your ability to solve problems in that domain. For most of this book we assume that the human who is representing a domain decides on the ontology and the relationships. To get human-level computational intelligence it must be the agent itself that decides how to divide up the world, and which relationships to reason about. However, it is important for you to understand what knowledge is required for a task before you can expect to build a computer to learn or introspect about how to solve a problem. For this reason, we concentrate on what it takes to solve a problem. It should not be thought that the problem of CI is solved. We have only just begun this endeavor.

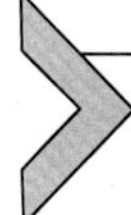

8. What is the impact of computational intelligence on society?

As with most changes in life, there will be positive and negative impacts on society as Computational intelligence continues to transform the world we live in. How that will balance out is anyone's guess and up for much debate and for many people to contemplate. As an optimist at heart, I believe the changes will mostly be good but could be challenging for some. Here are some of the challenges that might be faced (and we should be thinking about how to address them now) as well as several of the positive impacts Computational intelligence will have on society.

9. Challenges to be faced

The transformative impact of Computational intelligence on our society will have far-reaching economic, legal, political and regulatory implications that we need to be discussing and preparing for. Determining who is at fault if an autonomous vehicle hurts a pedestrian or how to manage a global autonomous arms race are just a couple of examples of the challenges to be faced.

Will machines become super-intelligent and will humans eventually lose control? While there is debate around how likely this scenario will be we do know that there are always unforeseen consequences when new technology is introduced. Those unintended outcomes of Computational intelligence will likely challenge us all.

Another issue is ensuring that CI doesn't become so proficient at doing the job it was designed to do that it crosses over ethical or legal boundaries. While the original intent and goal of the CI is to benefit humanity, if it chooses to go about achieving the desired goal in a destructive (yet efficient way) it would negatively impact society. The CI algorithms must be built to align with the overarching goals of humans.

Computational intelligence algorithms are powered by data. As more and more data is collected about every single minute of every person's day, our privacy gets compromised.

10. Positive impacts of computational intelligence on society

Computational intelligence can dramatically improve the efficiencies of our workplaces and can augment the work humans can do. When CI takes over repetitive or dangerous tasks, it frees up the human workforce to do work they are better equipped for—tasks that involve creativity and empathy among others. If people are doing work that is more engaging for them, it could increase happiness and job satisfaction.

With better monitoring and diagnostic capabilities, Computational intelligence can dramatically influence healthcare. By improving the operations of healthcare facilities and medical organizations, CI can reduce operating costs and save money.

One estimate from McKinsey predicts big data could save medicine and pharma up to $100B annually. The true impact will be in the care of patients. Potential for personalized treatment plans and drug protocols as well as giving providers better access to information across medical facilities to help inform patient care will be life-changing.

Our society will gain countless hours of productivity with just the introduction of autonomous transportation and CI influencing our traffic congestion issues not to mention the other ways it will improve on-the-job productivity. Freed up from stressful commutes, humans will be able to spend their time in a variety of other ways.

The way we uncover criminal activity and solve crimes will be enhanced with artificial intelligence. Facial recognition technology is becoming just as common as fingerprints. The use of CI in the justice system also presents many opportunities to figure out how to effectively use the technology without crossing an individual's privacy.

Unless you choose to live remotely and never plan to interact with the modern world, your life will be significantly impacted by artificial intelligence. While there will be many learning experiences and challenges to be faced as the technology rolls out into new applications, the expectation will be that Computational intelligence will generally have a more positive than negative impact on society.

11. Mapping CI use cases to domains of social good

For the purposes of this research, we defined CI as deep learning. We grouped use cases into 10 social-impact domains based on taxonomies in use among social-sector organizations, such as the CI for Good Foundation and the World Bank. Each use case highlights a type of meaningful problem that can be solved by one or more CI capability. The cost of human suffering, and the value of alleviating it, are impossible to gauge and compare. Nonetheless, employing usage frequency as a proxy, we measure the potential impact of different CI capabilities.

For about one-third of the use cases in our library, we identified an actual CI deployment (Fig. 1). Since many of these solutions are small test cases to determine feasibility, their functionality and scope of deployment often suggest that additional potential could be captured. For three-quarters of our use cases, we have seen solutions deployed that use some level of

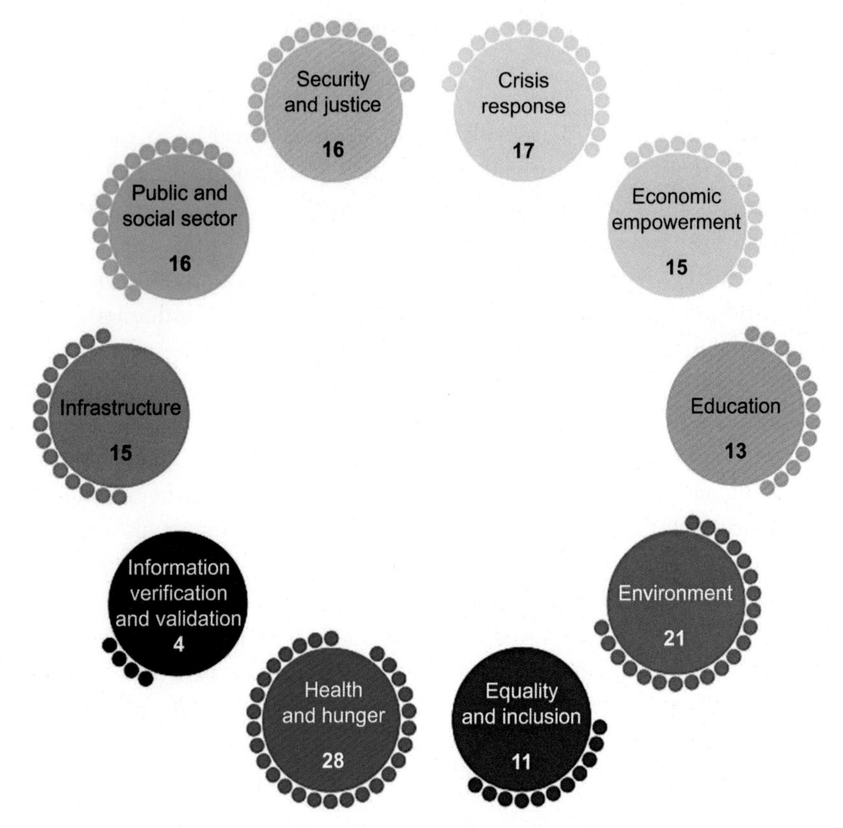

Fig. 1 Use cases that have impact on applying Computational Intelligence.

advanced analytics; most of these use cases, although not all, would further benefit from the use of CI techniques. Our library is not exhaustive and continues to evolve, along with the capabilities of AI.

11.1 Crisis response

These are specific crisis-related challenges, such as responses to natural and human-made disasters in search and rescue missions, as well as the outbreak of disease. Examples include using CI on satellite data to map and predict the progression of wildfires and thereby optimize the response of firefighters. Drones with CI capabilities can also be used to find missing persons in wilderness areas.

11.2 Economic empowerment

With an emphasis on currently vulnerable populations, these domains involve opening access to economic resources and opportunities, including jobs, the development of skills, and market information. For example, CI can be used to detect plant damage early through low-altitude sensors, including smartphones and drones, to improve yields for small farms.

11.3 Educational challenges

These include maximizing student achievement and improving teachers' productivity. For example, adaptive-learning technology could base recommended content to students on past success and engagement with the material.

11.4 Environmental challenges

Sustaining biodiversity and combating the depletion of natural resources, pollution, and climate change are challenges in this domain. The Rainforest Connection, a Bay Area non profit, uses CI tools such as Google's TensorFlow in conservancy efforts across the world. Its platform can detect illegal logging in vulnerable forest areas by analyzing audio-sensor data (Fig. 2).

11.5 Equality and inclusion

Addressing challenges to equality, inclusion, and self-determination (such as reducing or eliminating bias based on race, sexual orientation, religion, citizenship, and disabilities) are issues in this domain. One use case, based on work by Affective, which was spun out of the MIT Media Lab, and Autism Glass, a Stanford research project, involves using CI to automate the recognition of emotions and to provide social cues to help individuals along the autism spectrum interact in social environments.

11.6 Health and hunger

This domain addresses health and hunger challenges, including early-stage diagnosis and optimized food distribution. Researchers at the University of Heidelberg and Stanford University have created a disease-detection CI system—using the visual diagnosis of natural images, such as images of skin lesions to determine if they are cancerous—that outperformed professional dermatologists. AI-enabled wearable devices can already detect people with potential early signs of diabetes with 85% accuracy by analyzing heart-rate sensor data. These devices, if sufficiently affordable, could help more than 400 million people around the world afflicted by the disease.

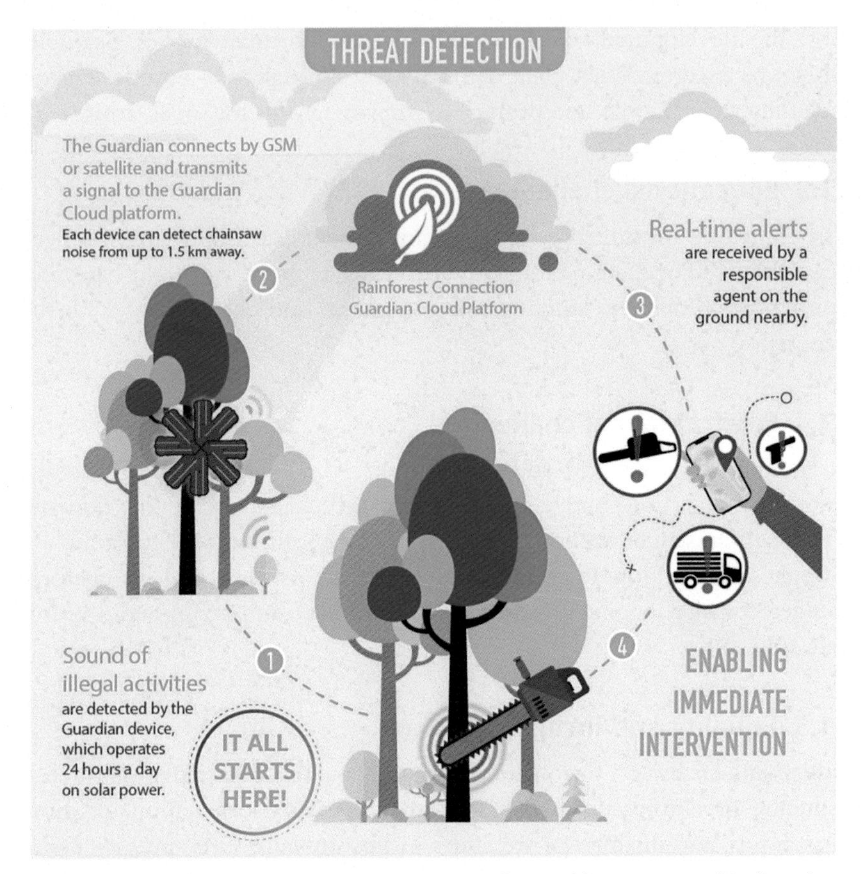

Fig. 2 Threat detection using Computational intelligence for sustaining biodiversity.

11.7 Information verification and validation

This domain concerns the challenge of facilitating the provision, validation, and recommendation of helpful, valuable, and reliable information to all. It focuses on filtering or counteracting misleading and distorted content, including false and polarizing information disseminated through the relatively new channels of the internet and social media. Such content can have severely negative consequences, including the manipulation of election results or even mob killings, in India and Mexico, triggered by the dissemination of false news via messaging applications. Use cases in this domain include actively presenting opposing views to ideologically isolated pockets in social media.

11.8 Public and social-sector management

Initiatives related to efficiency and the effective management of public- and social-sector entities, including strong institutions, transparency, and financial management, are included in this domain. For example, CI can be used to identify tax fraud using alternative data such as browsing data, retail data, or payments history.

11.9 Security and justice

This domain involves challenges in society such as preventing crime and other physical dangers, as well as tracking criminals and mitigating bias in police forces. It focuses on security, policing, and criminal-justice issues as a unique category, rather than as part of public-sector management. An example is using CI and data from IoT devices to create solutions that help firefighters determine safe paths through burning buildings.

12. Perceptions in depth

Till date, there is no universally accepted definition of intelligent computing. Some researchers regard intelligent computing as the combination of AI and computing technology. It marks 3 different milestones of intelligent computing systems according to the development of AI. This perspective limits the definition of intelligent computing within the field of AI while ignoring the inherent limitations of AI and the vital role of ternary interactions between humans, machines, and things. Another school of thought views intelligent computing as computational intelligence. This area imitates human or biological intelligence to realize optimal algorithms to solve specific problems and treats intelligent computing primarily as an algorithmic innovation. However, it fails to consider the essential roles that the computing architecture and the internet of things (IoT) play in intelligent computing.

We present a new definition of intelligent computing from the perspective of solving complex scientific and societal problems considering the increasingly tight fusion of 3 fundamental spaces of the world, i.e., human society space, physical space, and information space.

12.1 Definition 1

Intelligent computing is the area that encompasses the new computing theoretical methods, architecture systems, and technical capabilities in the era of digital civilization that supports the interconnection of all the world. Intelligent computing targets computational tasks with the minimum cost according to the specific actual needs, matching adequate computational power, invoking the finest algorithm, and obtaining optimal results.

The new definition of intelligent computing is proposed in response to the fast-growing computing needs of the triple integration of human society, the physical world, and information space. Intelligent computing is human oriented and pursues high computing capability, energy efficiency, intelligence, and security. Its goal is to provide universal, efficient, secure, autonomous, reliable, and transparent computing services to support large-scale and complex computational tasks. Fig. 3 shows the overall theoretical framework of intelligent computing, which embodies a wide variety of computing paradigms in support of human–physics–information integration.

First, computational intelligence is neither substitution nor a simple integration of the existing supercomputing, cloud computing, edge computing, and other computing technologies such as neuromorphic computing, optoelectronic computing, and quantum computing. Instead, it is a form of computing that solves practical problems by optimizing existing computing methods and resources systematically and holistically according to task requirements. In comparison, the major existing computing disciplines, such as supercomputing, cloud computing, and edge computing, fall into different domains. Supercomputing aims to achieve high computing power, cloud computing emphasizes cross-platform/device convenience and edge computing pursues quality of service and transmission efficiency. Intelligent computing dynamically coordinates the data storage, communication, and computation among edge computing, cloud computing, and supercomputing domains. It constructs various cross-domain intelligent computing systems to support end-to-end cloud collaboration, inter-cloud collaboration, and supercomputing interconnection.

Computational intelligence should make good use of existing computing technologies and, more importantly, promote the formation of new intelligent computing theories, architectures, algorithms, and systems. Second, computational intelligence is proposed to address problems in the future development of human-physics-information space integration. With the development of information technology applications in the big data era, the boundaries between physical space, digital space, and human society have become increasingly blurred. The human world has evolved into a new space characterized by the tight fusion of humans, machines, and things. Our social system, information systems, and physical environment constitute

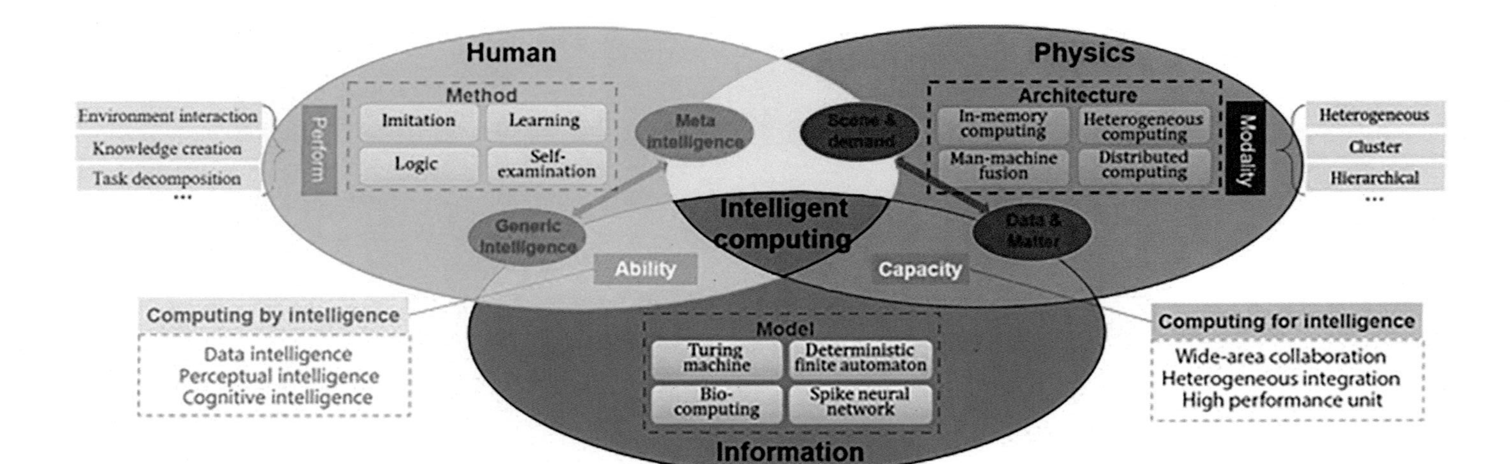

Fig. 3 Computational intelligence based on fusing human space, physical space and Information space.

a large dynamically-coupled system in which humans, machines, and things are integrated and interact in a highly complex manner, which promotes the development and innovations of new computing technologies and application scenarios in the future.

12.2 Conclusion

Computational intelligence has enormous potential to serve society, bringing more radical innovations for humans in the future. Its problem-solving ability could help people and communities around the world by solving today's toughest challenges. With sensible use of AI, we should continue to see a wide scope of AI applications and new developments for social good. Significant deployments have already changed decision-making, business models, risk mitigation, and system performance in banking, national security, health care, criminal justice, transportation, and smart cities. These changes are bringing significant economic and social benefits.

Further reading

[1] A.N. Shadowen, Ethics and Bias in Machine Learning: A Technical Study of What Makes us "Good", City University of New York, 2017.

[2] C. Corinne, Governing artificial intelligence: ethical, legal and technical opportunities and challenges, Philos. Trans. R. Soc. A Math. Phys. Eng. Sci. 376 (2018) 20180080.

[3] D. Kaufman, Inteligência Artificial e os desafios éticos, Paulus Rev. Comun. da Fapcom 5 (2021) 9.

[4] A. Hagerty, I, Rubinov, Global AI Ethics: A Review of the Social Impacts and Ethical Implications of Artificial Intelligence. White Paper, 2019.

[5] L. Vesnic-Alujevic, S. Nascimento, A. Pólvora, Societal and ethical impacts of artificial intelligence: critical notes on European policy frameworks, Telecomm. Policy 44 (6) (2020) 101961.

[6] https://www.nesta.org.uk/blog/using-artificial-intelligence-for-social-impact/.

[7] https://medium.com/@180dc.hansraj/the-social-impact-of-artificial-intelligence8e218cab7895.

[8] J. Morley, L. Floridi, L. Kinsey, A. Elhalal, From what to how: an initial review of publicly available ai ethics tools, methods and research to translate principles into practices, Sci. Eng. Ethics 26 (4) (2019) 2141–2168.

[9] I. Munoko, H.L. Brown-Liburd, M. Vasarhelyi, The ethical implications of using artificial intelligence in auditing, J. Bus. Ethics 167 (2) (2020) 209–234.

[10] L. Floridi, et al., AI4People—an ethical framework for a good AI society: opportunities, risks, principles, and recommendations, Mind. Mach. 28 (4) (2018) 689–707.

[11] T. Krupiy, A vulnerability analysis: theorising the impact of artificial intelligence decision-making processes on individuals, society and human diversity from a social justice perspective, Comput. Law Secur. Rev. 38 (2020) 105429.

[12] S.S. ÓhÉigeartaigh, J. Whittlestone, Y. Liu, Y. Zeng, Z. Liu, Overcoming barriers to cross-cultural cooperation in AI ethics and governance, Philos. Technol. 33 (4) (2020) 571–593.

[13] U. Pagallo, When morals ain't enough: robots, ethics, and the rules of the law, Mind. Mach. 27 (4) (2017) 625–638.

About the authors

Preetha Evangeline David is currently working as an Associate Professor and Head of the Department in the Department of Artificial Intelligence and Machine Learning at Chennai Institute of Technology, Chennai, India. She holds a PhD from Anna University, Chennai in the area of Cloud Computing. She has published many research papers and Patents focusing on Artificial Intelligence, Digital Twin Technology, High Performance Computing, Computational Intelligence and Data Structures. She is currently working on Multi-disciplinary areas in collaboration with other technologies to solve socially relevant challenges and provide solutions to human problems.

Anandhakumar is a professor in the Department of Information Technology at Anna University, Chennai. He has completed his doctorate in the year 2006 from Anna University. He has produced 17 PhD's in the field of Image Processing, Cloud Computing, Multimedia Technology and Machine Learning. His ongoing research lies in the field of Digital Twin Technology, Machine Learning and Artificial Intelligence. He has published more than 150 papers indexed in SCI, SCOPUS, WOS, etc.

CHAPTER TWO

Smart crisis management system for road accidents based on Modified Convolutional Neural Networks–Particle Swarm Optimization hybrid algorithm

V. Cephas Paul Edward
Anna University, Chennai, Tamil Nadu, India

Contents

Abstract

Statistics and reports show that road accident occurrence rates have increased significantly in the recent past. Machine learning and pattern matching techniques can be employed to manage road accidents in a smart manner with the intention to reduce the fatality rate and the damage incurred by the victims. The more the delay in access to medical help in case of more severe accidents, the higher the fatality rate and also increased risk of complications. Leveraging the surveillance infrastructure and implementing smart algorithms on the basis of crash detection along with the estimation of the speed of the vehicle, trajectory analysis and also trying to come up algorithmically with the impact would be one side of the crisis management system. The proposal involves employing a modified CNN (Convolutional Neural Networks) and Particle Swarm Optimization hybrid for vehicle trajectory analysis and estimating

Advances in Computers, Volume 132
ISSN 0065-2458
https://doi.org/10.1016/bs.adcom.2023.07.002

the impact and arriving at the severity of the accident. The approach will be applicable to various accident scenarios like inter-vehicle collision, vehicle-human collision, vehicle self-accidents, etc.

The other equally important will be to alert the nearest medical emergency center based on the estimated impact will ensure ambulances with all required assistance. Apart from the first response, other actions could include recommended alternate routes for navigation, suggestions of hospitals depending on the estimated impact, sending out notifications to emergency contacts and informing the local cops/ insurance agencies, etc. The article also proposes an optimal model based on geo-spatial particle swarm optimization for the above problem statement.

1. Background

Statistics and reports show that road accident occurrence rates have increased significantly in the recent past. In over 80% of these accidents, a timely crisis management hadn't been provided and hence resulting in mishaps. Timely crisis management includes detecting the accident on time and providing the required emergency care to the victim(s) on time [1]. Studies also show that the rate of causality will increase to 8% approximately by 2030 from the 4% rate that prevails currently. More than 50% of all road traffic deaths are of pedestrians, cyclists, and motorcyclists and affecting the age bracket of 5–29 years [2]. There are many studies to detect vehicular crashes based on dashboard sensors and other devices such as GPS, accelerometer and linear acceleration sensors. However, this study is limited to accident detection system from a third-person perspective, i.e., street/traffic cameras. As mentioned before, the more the delay in access to medical help in case of more severe accidents, the higher the fatality rate and also increased risk of complications. Leveraging the existing traffic surveillance infrastructure is a good option to implement smart crash detection algorithms involving the estimation of the speed of the vehicle, trajectory analysis and also trying to come up algorithmically with the impact. Notification mechanism to relevant agencies like medical help, law enforcement, fire station and emergency contacts etc. would also need to be performed quickly for an effective crisis handling.

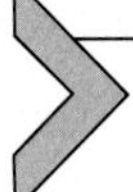

2. Implementation

2.1 Overview

In order to estimate the impact of the collision or crash, data points such as velocity of the vehicle (if the accident was a vehicle colliding with a stationary object like a median) or relative velocity of the vehicle in case of two vehicles colliding against each other. Additionally, trajectory analysis also provides us added value to the impact analysis.

2.2 Moving vehicle detection

There are various approaches that can be employed to detect and assess the changes in moving vehicles.

- Image thresholding: The overall objective is to obtain a good binary image after removing all irrelevant areas. As a pre-requisite, the image needs to be converted into grayscale. There is a fixed threshold that is agreed upon and each pixel is compared with the threshold. Depending on the comparison outcome, it is assigned one of two pre-defined values. The simplest form of thresholding is the binary thresholding.

$$p'(x, y) = \begin{cases} 0, & \text{if } p(x, y) \leq \text{threshold} \\ \text{max\_val}, & \text{if } p(x, y) > \text{threshold} \end{cases}$$

where p′ is the transformed image and p is the original image with x,y representing each pixel. Inverted Binary thresholding can also be used for the purposes of moving vehicle detection (Fig. 1).

$$p'(x, y) = \begin{cases} 0, & \text{if } p(x, y) > \text{threshold} \\ \text{max\_val}, & \text{if } p(x, y) \leq \text{threshold} \end{cases}$$

- Frame differencing: A video is a continuous stream of image frames. A moving vehicle implies that the vehicle changes its position across the frames while the other portions of the frames are relatively static. Thus, comparing two versions of frames across time can be leveraged for detecting moving vehicles (Fig. 2).
- Contour based: This approach is based on "regions of interest." Shapes of the objects are found by setting boundaries around region of interest and co-ordinates of the contours can be used to detect the moving vehicle (Fig. 3).

Fig. 1 (A) Original image; (B) Binary thresholding; (C) Inverted Binary thresholding.

Fig. 2 Moving vehicle detection with frame differencing.

Fig. 3 Moving vehicle detection with Contour-based approach.

2.3 Velocity calculation

In general, speed is calculated as:

$$speed = \frac{distance}{time}$$

To implement this on top of the vehicle detection algorithm involves calibration of the output of the vide camera to an actual distance, i.e., a polynomial area is chosen on the vide frames and attributed to the "real world" distance [3–5]. For example in Fig. 4, the distance represented by the yellow arrow has to be derived from the real world distance to camera output ratio (Fig. 4).

Fig. 4 Calibration of "real world" distance.

Ideally, the proposed approach eliminates the need for this calibration by considering the relative distances covered by objects between two consecutive frames. But for velocity calculation and vehicle trajectory analysis, object tracking is required. It is defined as the process of maintaining the state of an object as it moves to newer positions. Object tracking broadly involves:

(1) As input: detected vehicles as part of the previous step
(2) Assignment of unique ID to each "new" detection
(3) Tracking each object with the ID

There are various algorithms for object tracking. The current proposal utilizes "Centroid-based" object tracking (Fig. 5).

The centroid-based tracking algorithm expects an array of "bounding" rectangles or boxes for each detected vehicle across all frames. This can be obtained by Color thresholding + Contour extraction as above or HAAR cascades, R-CNNs, SVMs, etc. For each box, the center point is then calculated. The next step of the algorithm proceeds with computation of Euclidean distance between the center points (also known as "centroids") (Fig. 6). Closer the distance, the objects are the same. This approach is used to identify any new objects detected and assign them with new IDs. The existing centroid also need to be updated with the change in frame to account for the object movement. The algorithm also should have a provision to drop any object that is disappeared from the field of view [6].

Further accuracy is improved by employing algorithms such as DEEP sort that works based on YOLO object detection.

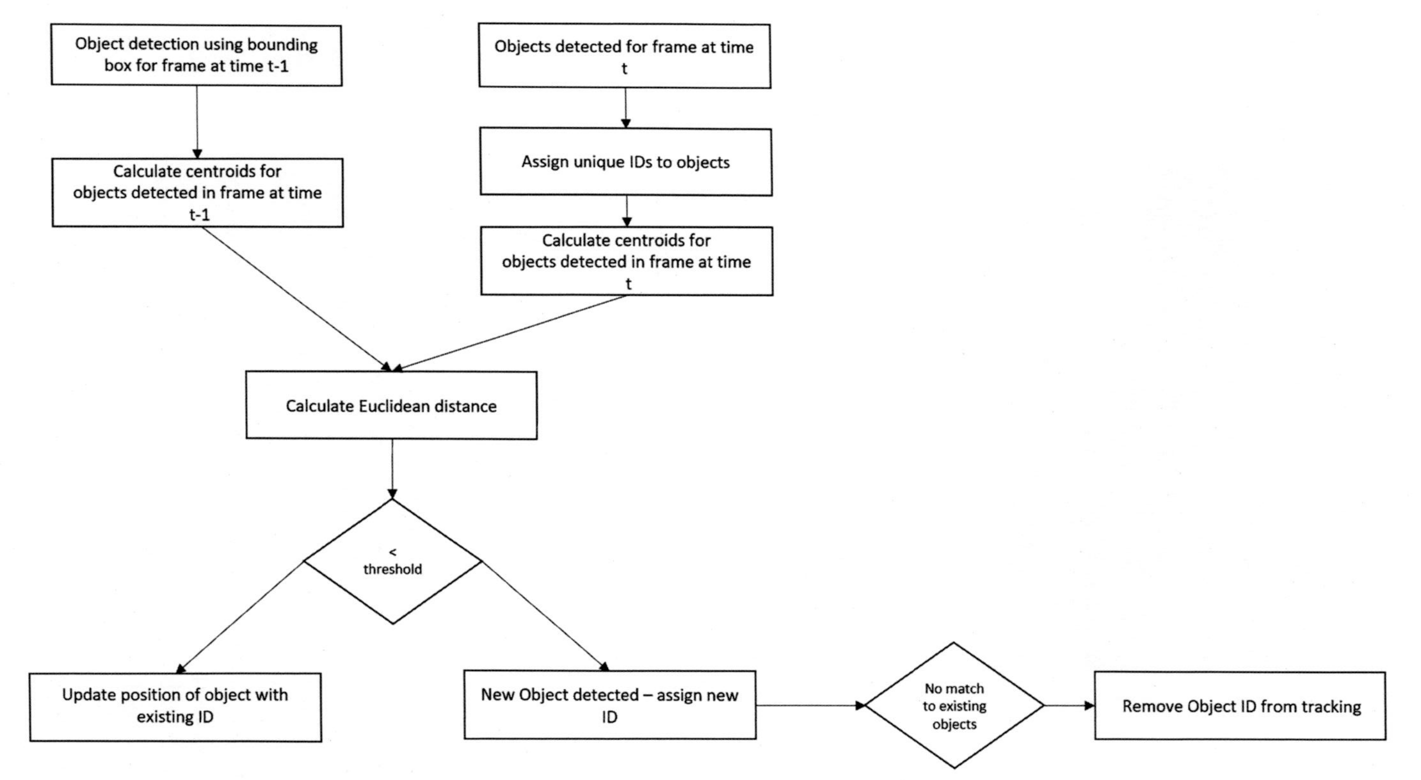

Fig. 5 Centroid-based object tracking flowchart.

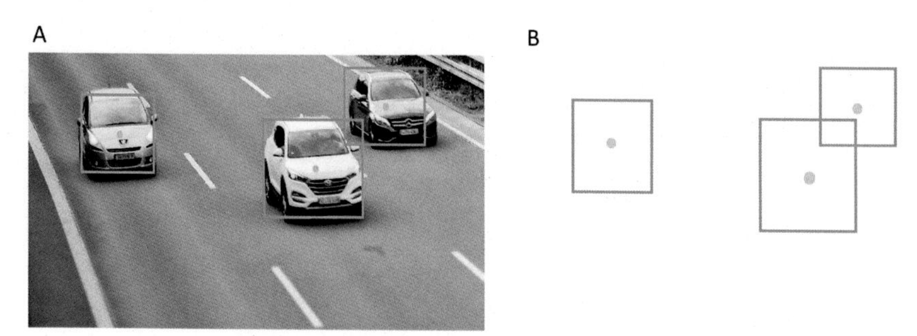

Fig. 6 (A) Centroid-based object tracking. (B) Two vehicles identified with bounding boxes.

Once object tracking is in place, the change in position of the objects are processed to fetch the "speed" equivalent which is computed as follows:

$$\frac{\sum p_t(x', y') - p_{t-1}(x, y)}{\sum n(frames)}$$

This is computed for each detected object. Along with the "speed," the other important feature required for severity analysis is the trajectory.

2.4 Vehicle trajectory analysis

For each object identified, the change in position across frames is tracked and plotted as a polyline to arrive at the object trajectory. An approach similar to the one described in Ref. [7] is followed where both the distances between objects and the vehicle trajectories are taken into account. On top of this, the speed data is overlayed, i.e., for every point in the trajectory that is obtained via frame differencing (not the ones extrapolated using polyline construction) (Fig. 7). CNN is used to extract features, i.e., differentiate the input trajectories along with the in-point speeds and also assign learnable weights and CNN biases. Images were converted into standard size of 256×256 and processed through a convolutional filter of 3×3. These images which are smaller in size are used to create feature maps. Max pooling to down sample multiple feature maps is applied to reduce the dimensionality of the feature map extracted. Rectified linear units (RELU) is chosen as the activation function.

2.5 Accident detection and severity quotient

In addition to the trajectories and speed, the size of the moving objects is something that needs to be taken into account during the impact prediction.

Fig. 7 (A) Image segmentation. (B) Object identification and tracking. (C) Trajectory plotted.

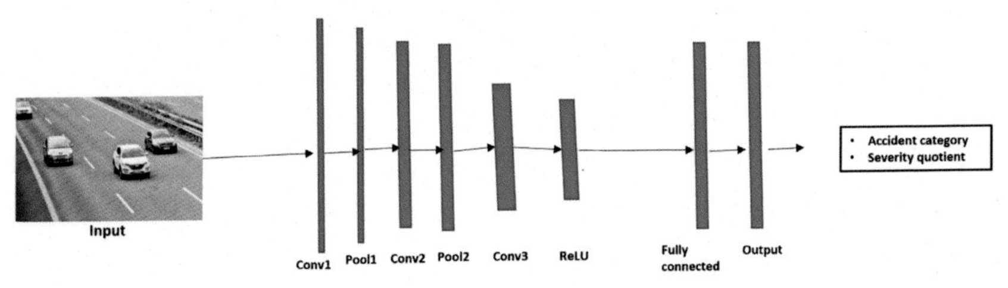

Fig. 8 Accident severity detection and classification CNN.

Neural networks in general are suited for classification and pattern recognition problems. The CNN/ConvNet-based approach is used to detect the occurrence of an accident of collision. Furthermore, the accident can be classified as inter-vehicle collision, vehicle-human collision, vehicle self-accidents (vehicles colliding with medians, trees or similar stationary objects) [8]. The severity score needs to be adjusted based on the type of accident. The proposed method refers to the severity impact in terms of "severity quotient" after taking into account the category accidents as mentioned above. Particle Swarm Optimization is chosen further enhance and arrive at accident category and severity quotient. CNN is a feed forward, deep neural network which is able to differentiate several input images and assign learnable weights/biases to various components of an image. This is very much inspired from the Visual Cortex and resemble the connectivity pattern of neurons in the human brain (Fig. 8) [9,10].

- The core of the CNN construct is the convolution layers. The idea of convolution layers is that, the input image at each layer will activate a particular characteristic of the image. The output of one layer feeds into the other layer. The size of the feature map extracted is inversely proportional to the number of convolutional layers employed.
- Rectified Linear unit layer is also applied to remove any negative values by resetting them to zero values in order to improve the efficiency.
- Pooling layer helps in down sampling the width and height of the frame and arrive at fewer parameters.
- The model used for arriving at severity quotient and for accident classification consists of three convolutional layers and with Max pooling and one ReLU layer.

Additionally, back propagation is used to tweak the core weights which is an optimization problem. Various hyper-parameters are kernel length, stride, dropout rate, max-pooling kernel length, etc. This is where Particle Swarm Optimization can be blended in. The algorithm has been inspired from nature, specifically fishes or birds grouping together. Each data point is referred to as a particle and resembles a fish trying to search for food. A leader fish spots the food and the entire group of fishes follow to get hold of the food. The same principle is applied for global optimization problems [11].

Consider the Optimization function "f," initially an agent population (particle population) needs to be created that follows uniform distribution. Let this population of agents be referred to as "PA." Each particle's position needs to be evaluated in context of "f." If the present position of the agent/particle is found to be better than then previous value, then the value is updated. Find the best particle and update the velocities.

$$V^{t+1} = W.V^t + C_c.U_1^t\left(P_b^t - P^t\right) + C_s.U_2^t\left(G_b^t - P^t\right)$$

where,

V^t is the velocity of a particle at time instant "t"
W is the inertia weight
C_c is the Cognitive constant
C_s is the Social constant
P_b^t is the personal best of a particle at time instant "t"
G_b^t is the global best of a particle at time instant "t"
U_1, U_2 are random numbers
Move the particles to new positions: $P^{t+1} = P^t + V^{t+1}.$

Table 1 Sample accident severity quotient and category data.

Size of blob	Speed equivalent	Trajectory	Severity quotient (normalized)	Accident category
24	12		6.1	Vehicle–vehicle (same direction)
5	19		5.9	Vehicle–vehicle (opposite direction)
12	32		8.2	Vehicle–object collision

Keep iterating over this until particle position satisfies the criteria of the objective function. The severity quotients are normalized on a scale of 1–10 (Table 1).

Based on the severity quotient, the required medical equipment, etc. can be summoned for as part of the ambulance dispatch algorithm.

3. Post-detection steps

Once the accident has been detected, it is equally important to alert the nearest medical emergency center based on the estimated impact will ensure ambulances with all required assistance. Apart from the first response, other actions could include recommended alternate routes for navigation, suggestions of hospitals depending on the estimated impact, sending out notifications to emergency contacts and informing the local cops/insurance agencies, etc. The article also proposes an optimal model based on geospatial machine learning for the above problem statement.

In Ref. [12], the authors have described an approach based on reinforcement learning to solve the issue of routing the appropriate ambulance to the accident scene. The approach is based on Multi-Agent Q-Network with Experience Replay(MAQR) and they have stated that studies on ambulance dispatch are not flexible enough to adapt to dynamic situations as they rely on simple spatial data.

Initial dataset is the geospatial distribution of hospitals within the area along with the corresponding ambulance locations and the accident locations (Fig. 9).

Following table represents some sample data of the post-detection problem specifically with respect to ambulance/hospital. Refs. [9,10,13–15] discuss various approaches around the ambulance dispatch problem. But the

Fig. 9 Geospatial distribution of accidents and ambulance data.

Table 2 Sample data on hospital, ambulance availability vs accident severity.

Accident severity	Hospital	Hospital speciality	Hospital proximity	Ambulance	Ambulance proximity	Ambulance availability
5.9	H1	4	2	A1	3	True
6.1	H1	4	5	A2	3	False
8.2	H1	4	7	A3	2	True
5.9	H2	7	8	A4	5	True
6.1	H2	7	3	A5	6	True
8.2	H2	7	6	A6	2	False
5.9	H3	9	2	A7	9	True
6.1	H3	9	2	A8	1	True
8.2	H3	9	3	A9	7	False

proposed approach varies in terms of the parameters chosen and the methodology followed.

Hospital speciality is a score that is arrived based on the facilities, staffing and treatment success rate of a particular hospital. This needs to match up with the severity quotient of the accident such that complex cases are routed to hospitals with higher speciality ratings. Particle swarm optimization can also be applied on top of the modified geospatial data as mentioned the Table 2. This needs to be weighted against additional parameters like traffic volume, junction delay, road width, road type, travel time, etc. to come up with the most optimal (global best) ambulance/hospital combination for a specific accident. Sending out notifications to local cops, automated update of accident event on navigation systems like maps are done on top of the severity quotient computation. Additionally, once the ambulance dispatch optimization is performed and the ambulance along with the route is determined, other vehicles/commuters can be notified via their navigation systems (maps) to be cautious and aware of ambulance's route and perhaps pick alternatives. The cops can also be notified with this information.

Sending out notifications to emergency contacts and insurance agencies can be performed once the emergency medical services arrive and provide the necessary first aid/care.

4. Conclusion

The proposed approach takes into account the fact that more the delay in access to medical help in case of more severe accidents, the higher the fatality rate and also increased risk of complications. The approach involved estimation of the size and speed of the vehicle, trajectory analysis employing a modified CNN (Convolutional Neural Networks) and Particle Swarm Optimization hybrid for vehicle trajectory analysis and estimating the impact and arriving at the severity of the accident. The approach will be applicable to various accident scenarios like inter-vehicle collision, vehicle-human collision, vehicle self-accidents, etc. Particle Swarm Optimization is also applied to geo-spatial data for emergency service/ambulance dispatch. Currently, the experiment was carried out based on the Highway Traffic Videos Dataset [16] from a third party perspective camera. But eventually this can be extended to various other datasets and also similar experiment can be conducted form a First party perspective camera or dashboard cam.

References

[1] M.S. Supriya, S.P. Shankar, B.J. Himanshu Jain, L.L. Narayana, N. Gumalla, Car crash detection system using machine learning and deep learning algorithm, in: *2022 IEEE International Conference on Data Science and Information System (ICDSIS)*, 2022.

[2] Shweta, J. Yadav, K. Batra, A.K. Goel, A framework for analyzing road accidents using machine learning paradigms, J. Phys. Conf. Ser. (2021), https://doi.org/10.1088/1742-6596/1950/1/012072. *1950 (2021) 012072 IOP Publishing*.

[3] M.D.M. Khan, K. Srinivas, J. Kavitha, Vehicle speed detection using Python, Int. Res. J. Eng. Tech. 7 (5) (2020) 240–245.

[4] H. Dwivedi, S. Jain, A. Garg, Dr. Achal Kaushik, Speed detection software, Int. J. Eng. Res. Tech. 10 (3) (2021) 599–607.

[5] N.K. Budhnerm, K. Rangam, S. Shinde, S. Sayyed, Z. Tamboli, Vehicle speed detection using open CV in python, Int. Res. J. Mod. Eng. Technol. Sci. 5 (2) (2023) 2443–2446.

[6] B. Jacob, E. Violette, Vehicle trajectory analysis: an advanced tool for road safety, Procedia, Soc. Behav. Sci. 48 (2012) 1805–1814.

[7] Y. Zhang, Y. Sung, Traffic Accident Detection Method Using Trajectory Tracking and Influence Maps, MDPI Academic Open Access publishing, Mathematics, 2023.
[8] A. Khosroshahi, E. Ohn-Bar, M.M. Trivedi, Surround vehicles trajectory analysis with recurrent neural networks, in: 2016 IEEE 19th International Conference on Intelligent Transportation Systems (ITSC), 26 December, 2016.
[9] B. Kubeda, F. Zhang, A. Oluwasanmi, F. Owusu, M. Assefa, T. Amenu, Vehicle Accident and Traffic Classification Using Deep Convolutional Neural Networks, IEEE, 2020.
[10] N.H. Phong, A. Santos, B. Ribeiro, PSO-convolutional neural networks with heterogeneous learning rate, *IEEE* Access 4 (2016).
[11] M. Mythili, P. Pavithra, R.K. Selvakumar, Road accident vehicle detection using Particle Swarm Optimization technique, Int. J. Adv. Res. Comput. Sci. Eng. Inf. Technol. 4 (3, Special Issue: 2) (2016) 881–888.
[12] K. Liu, X. Li, C.C. Zou, H. Huang, F. Yanjie, Ambulance dispatch via deep reinforcement learning, in: *SIGSPATIAL '20: Proceedings of the 28th International Conference on Advances in Geographic Information Systems November, 2020.*
[13] S. Ahmed, R.F. Ibrahim, H.A. Hefny, An efficient ambulance routing system for emergency cases based on Dijkstra's algorithm, AHP, and GIS, in: *The 53rd Annual Conference on Statistics, Computer Science and Operation Research 3–5 Dec, 2018.*
[14] J.R. Challapalli, N. Devarakonda, A novel approach for optimization of convolution neural network with hybrid particle swarm and grey wolf algorithm for classification of Indian classical dances, Knowl Inf Syst. 64 (2022) 2411–2434.
[15] H. Hajari, M.R. Delavar, Particle swarm optimization in emergency services, international archives of the photogrammetry, Remote Sensing and Spatial Information Science XXXVIII (Part 8, Kyoto Japan) (2010) 326–329.
[16] Website: https://www.kaggle.com/datasets/aryashah2k/highway-traffic-videos-dataset?resource=download Highway Traffic Videos Dataset distributed under Public Domain license - https://creativecommons.org/publicdomain/zero/1.0/.

Further reading

[17] Website: https://www.analyticsvidhya.com/blog/2021/10/an-introduction-to-particle-swarm-optimization-algorithm/#:~:text=PSO%20is%20a%20stochastic%20optimization,looking%20for%20the%20best%20solution.
[18] Website: https://www.codingninjas.com/codestudio/library/vehicle-detection-using-opencv.
[19] Website: https://github.com/lev1khachatryan/Centroid-Based_Object_Tracking.
[20] Website: https://medium.com/aiguys/a-centroid-based-object-tracking-implementation-455021c2c997.

About the author

V. Cephas Paul Edward is a postgraduate in Computer Science and Engineering from Anna University MIT Campus, Chennai. He had completed his post-graduation in 2016 and is committed to contribute to research during and after his period of study. His areas of specialization include Image processing, DevOps including containerization and metrics.

CHAPTER THREE

Residential energy management system (REMS) using machine learning

G. Ramya[a] and R. Ramaprabha[b]

[a]Department of Computing Technologies, Faculty of Engineering and Technology, SRM Institute of Science and Technology, Kattankulathur, Chennai, India
[b]Department of EEE, SSN College of Engineering, Chennai, India

Contents

Abstract

Everywhere in the world, people are using more electricity every day, which causes a supply and demand imbalance. Everyone looks for more affordable and ecologically friendly ways to acquire electricity in the current situation. Energy management, a practical and popular strategy, might be used by the utility and the consumer to prevent a serious energy crisis. According to our opinion, it ought to begin at the consumer scope level. This study suggests a machine learning-based approach for creating a smart Residential Energy Management System (REMS). Utilizing machine learning algorithms, proposed method efficiently shifts early predicted priority loads between grid and rooftop renewable energy system at the residential buildings, without limiting consumption. By implementing load shifting algorithm, it is also possible to reduce the electricity expenditure in residential buildings while maintaining a steady power supply.

Advances in Computers, Volume 132
ISSN 0065-2458
https://doi.org/10.1016/bs.adcom.2023.07.003

The system also includes capabilities for predicting average solar power that is currently available using artificial neural networks and for maximizing the use of solar power generation and energy storage that is currently available using reinforcement learning. In the end, the Residential premises have less dependence on the grid. According to our research, small-scale home and residential industries should be the starting point for improving and promoting energy management. In order to provide a more consistent supply while lowering the electricity price, load shifting is demonstrated in this study at a residential building with solar powered sources. By machine learning methods, the suggested system (REMS) in this research efficiently switches potential loads between the local battery energy storage powered by renewable source and the local grid.

1. Introduction

Due to growing population the need for electrical sources has been increased. Conventional grid doesn't compensate the energy demand due to many issues. High utilization of fossil fuels and other non-renewable sources lead to environmental defects [1–4]. The demand can be rectified by reducing the grid dependency and increase the use of renewable resources. It is a key duty for all customers that facilitates to alleviate strength disaster and weather modifications to great extent. Energy management at source side depends on time of use (TOU) and control of load [2–6]. The above parameters are determined it is utilized at demand side consumers. Load side techniques are as follows power factor correction, peak clipping and valley filling. It results in greater dependable deliver to the customers with much less eco-friendly threats.

Load side management (DSM) is the main issue in controlling the demand from the consumers. This issue can be mitigated by using techniques like load shifting, load scheduling etc. In latest years, a number of the research work have targeted on Smart REMS in distinctive ways and gives numerous technical methodologies. In literature [5] reveals that smart EMS is carried out with the use of recent booming IoT technology. It gives the clarification to a scenario of load demand to vicinity even as matching the user demand such that, the proposed device substitutes with a fractional load losing system in a manageable method. Besides, it additionally reflects the client load choice in decision making of proposed system. Issues with the proposed structure is continuous energy to load demand at maximum, usage of renewable energy sources running with machine learning technique [7–11]. Wi-Fi communication is implemented for long distance communication.

In literature [2], an enhanced demand organization technique for EMU/HEMS incorporated with smart algorithm (ML) is proposed. It shares the loads between renewable and non-renewable sources in a grid tied environment. Research gap in the developed system is utilization of privately empowered battery capacity with rooftop sunlight based, accessibility of normal sun oriented power forecast utilizing ML and battery optimization strategy incorporated with Reinforcement Learning. In [6], Machine Learning based REMS that actually changes potential burdens to sustainably stimulated neighborhood energy capacity in light of its charge-release exchanges and matrix accessibility, accordingly decreasing the power utilization. Research hole is accessibility of normal sun based power forecast utilizing ML method and intelligent battery management system.

In our proposed system, it is focused to upgrade and elevate energy the executives should be begun from ground level. This paper presents load moving at a private reason with a nearby energy stockpiling empowered by roof sun oriented, utilizing machine learning algorithms to give a solid inventory however much as could be expected while diminishing the power bill.

2. Proposed smart energy management system

Proposed architecture in this research successfully change potential barrier between the principal lattice and sustainably empowered neighborhood battery capacity stimulated by housetop sunlight based at the private buildings by utilizing machine learning. The battery is charged exclusively by natural sunlight powered PV module [12–15]. As indicated by the charge-release exchanges of the battery and framework accessibility, pre-focused on list (from client) of burdens are exchanged and worked either with regular matrix power or/and non-traditional sustainable power (battery). This methodology attempts to supply source to loads however much as could reasonably be expected, without restricting utilization. Fig. 1 depicts the general model of the proposed system.

For our review, we consider domestic loads like heaters, lights, Fans, AC and Coolers. The data considered for this research work is collected from several residential load demand requirement and power consumption of various loads at residence is depicted in Fig. 2.

For research purpose loads like fans and AC are considered in majority in resistance. Lights are considered as low utilized sources during day time.

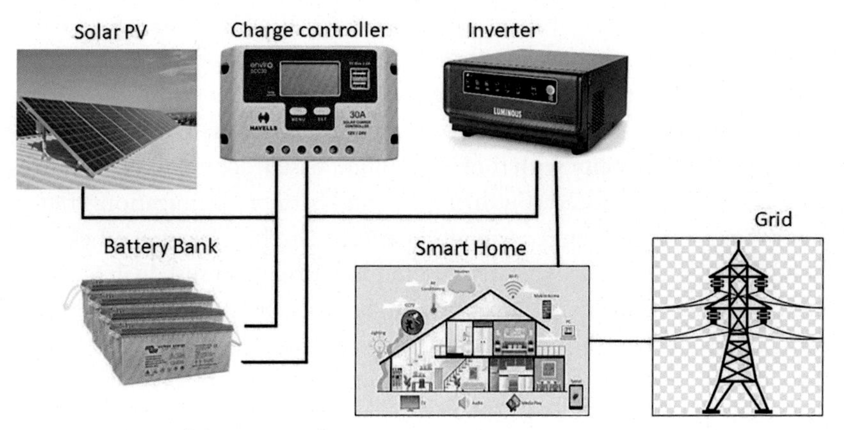

Fig. 1 Overview of the proposed system.

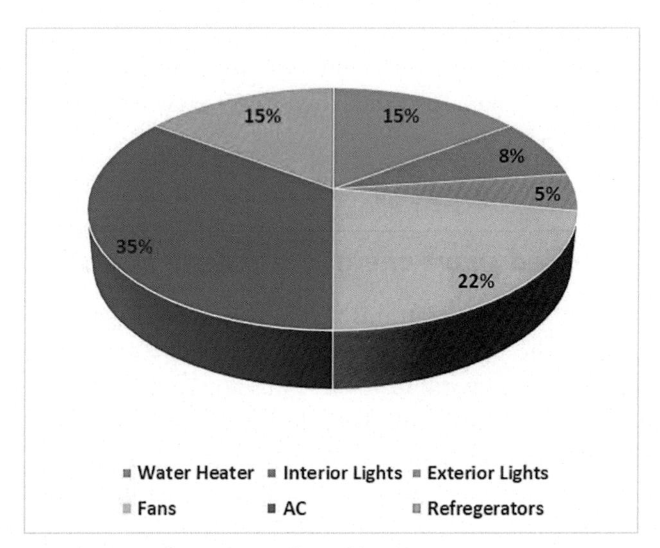

Fig. 2 Power utilization of various residential loads.

Initially load priority list to be ensured before the execution of proposed system. The overall proposed system is depicted in Fig. 3. The proposed system contains battery storage, solar system, grid, loads and controllers. Similar loads are segregated together and further all combined together which is controlled by centrally connected main controller.

The average utilization of solar sources is determined by using various sensors associated with Artificial Neural Network (ANN). The utilized power from PV source is used in the battery optimization algorithm which

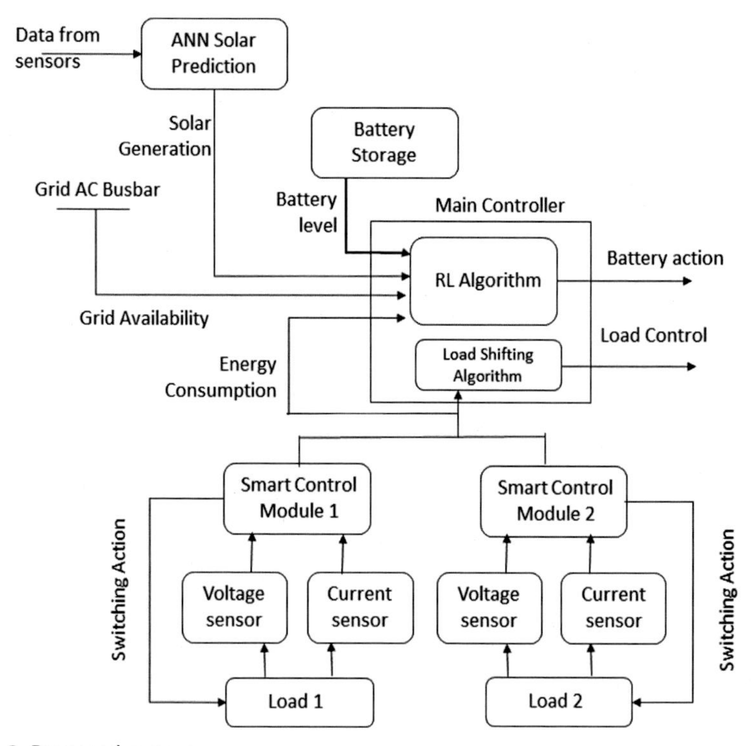

Fig. 3 Proposed system.

helps in controlling the main controller. The reinforcement learning methodology is implemented to run the main controller. Based on the algorithm the battery will either charge/discharge or remains idle. Based on the proposed system the loads are exchanged among main grid and battery as per load priority list.

3. Proposed smart REMS model

Proposed REMS model is discussed based on three situations in which two main conditions are discussed in detail.

Grid is accessible: Reinforcement Learning algorithm is used for battery optimization. The load shifting technique is implemented to decide the battery charging/discharging.

Grid is not accessible: at this condition the high priority loads are charged from the battery based on the charge availability in battery.

3.1 Case1: When the grid is available and the battery is charging

Load are utilized by the grid power. Battery gets charging by using solar source and the excess energy is fed back to grid. Solar system is linked to grid tied inverter. Here, the power flow is unidirectional such as the power doesn't flow from battery to grid or vice versa as shown in Fig. 4.

3.2 Case 2: When the grid is available and the battery is discharging

In this scenario the low utilized loads are energized by battery and grid. Generated power from solar are fed back to the grid and PV connected grid through inverter as shown in Fig. 5.

3.3 Case 3: When grid is not available

In this case high demand load is considered and motorized by the battery. Solar system is detached from the battery when the discharging starts as shown in Fig. 6.

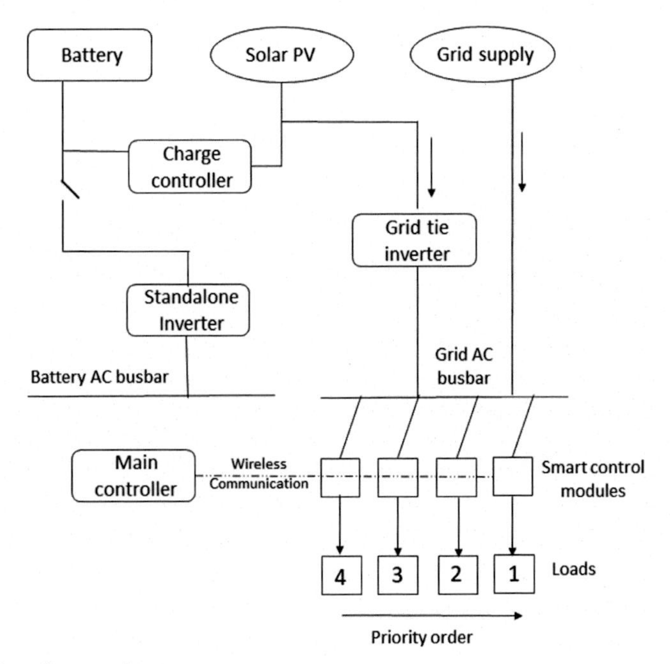

Fig. 4 Line diagram for case 1.

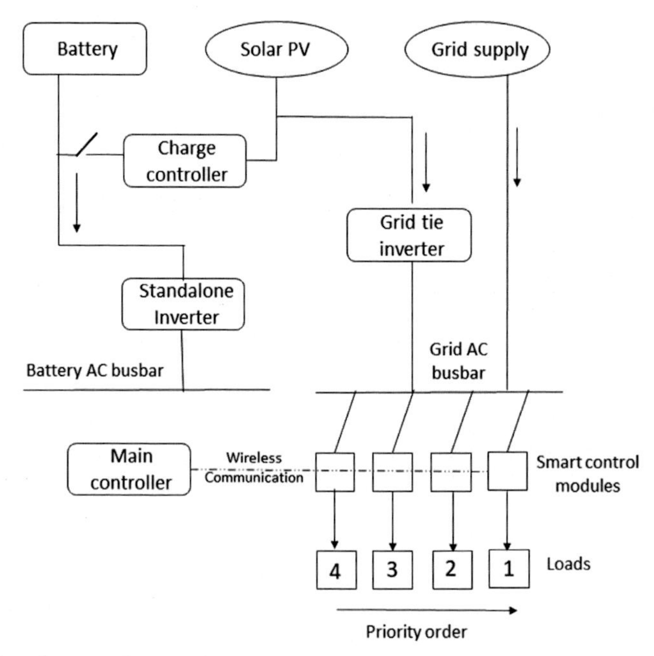

Fig. 5 Line diagram for case 2.

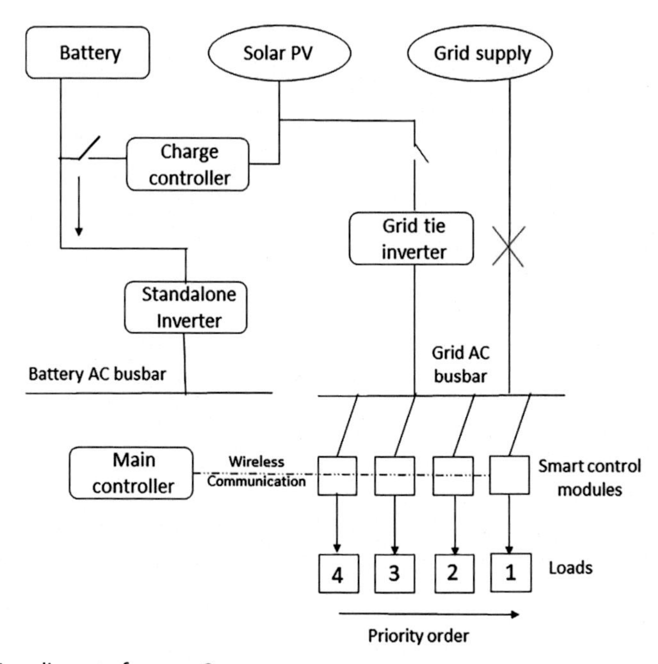

Fig. 6 Line diagram for case 3.

4. Load shifting algorithm

In this algorithm the load is supplied by grid and battery based on priority list by the user. The battery power P_B, Load Power P_i and Load priority order (1, 2, …, n) are considered when grid is available and not available.

4.1 Case 1: When the grid is available

In this case the battery functioning is determined by Reinforcement Learning (RL) algorithm. Initially when the battery is charging all loads are connected to grid. At this stage the solar PV system is connected to both grid and battery. During discharging of battery the less prioritized load is connected to battery. The flow chart is shown in Fig. 7.

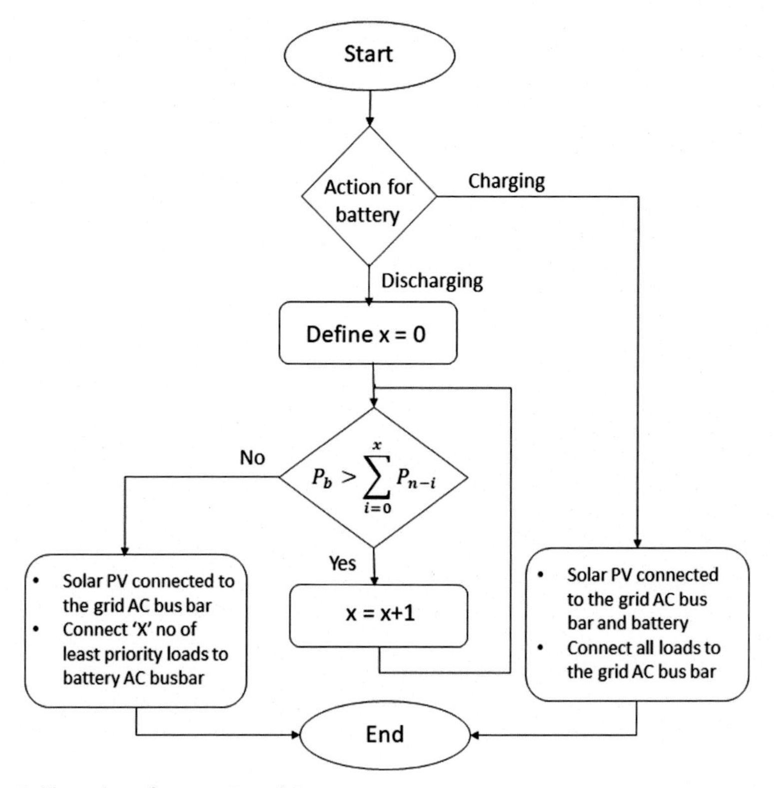

Fig. 7 Flow chart for case 1 and 2.

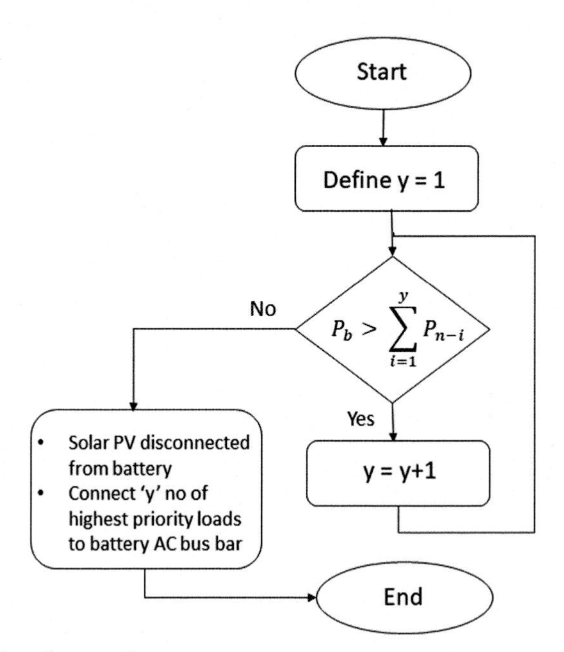

Fig. 8 Flow chart for case 3.

4.2 Case 2: When grid is not available

In this case when the grid is not available the high demand loads are operated based on the priority. At this stage the battery discharge is crucially considered and depicted in Fig. 8.

5. Solar power generation prediction using ANN

In this algorithm two hidden layers are considered. A supervised learning approach is used. An appropriate data for training, validating and testing the models are considered from National Renewable Energy Laboratory website.

Three different models were trained using ANN with various input features such as hour, wind speed, ambient temperature and plane of array irradiance as shown in Fig. 9. The root mean square error is determined and represented graphically presented in Fig. 10.

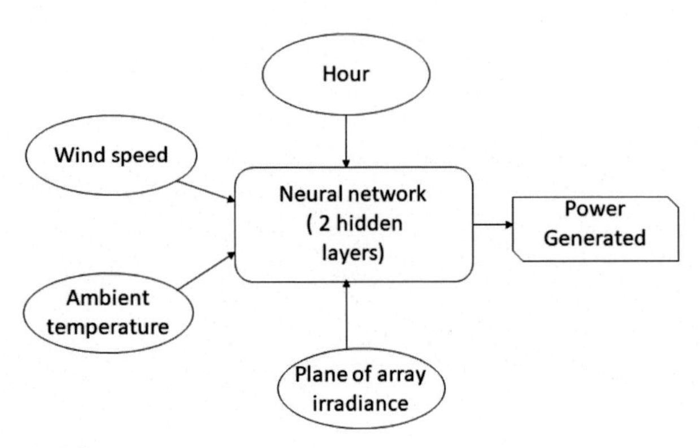

Fig. 9 ANN model.

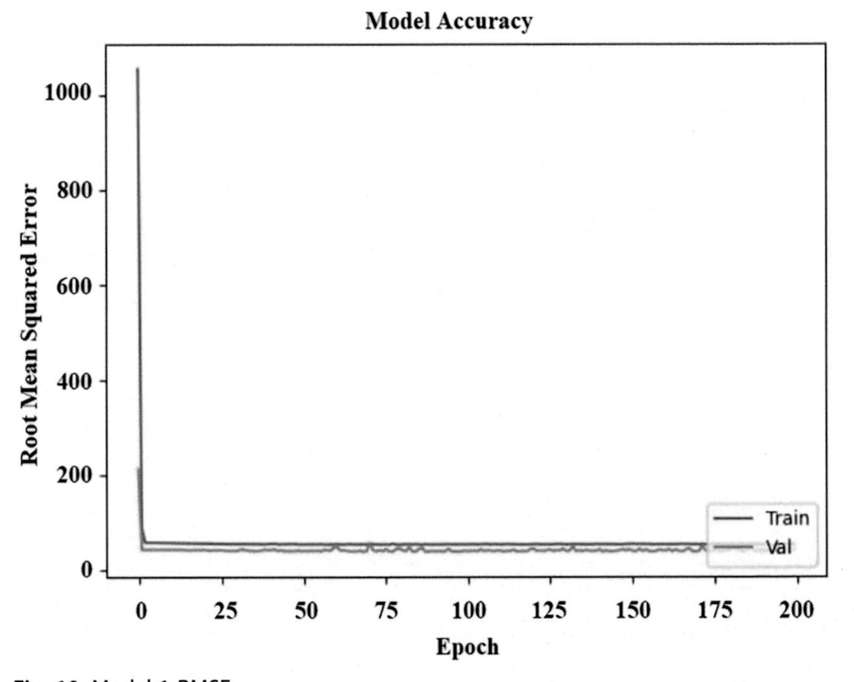

Fig. 10 Model 1 RMSE.

6. Battery optimization algorithm

This algorithm is carried out when grid is connected to the load. Reinforcement Learning algorithm is incorporated at this stage (Fig. 11). Machine learning technique is concerned with in what way the analyst will take action in an environment to amplify the idea of aggregate award [3].

Environmental conditions:

1. Foreseen solar power generation with time step t

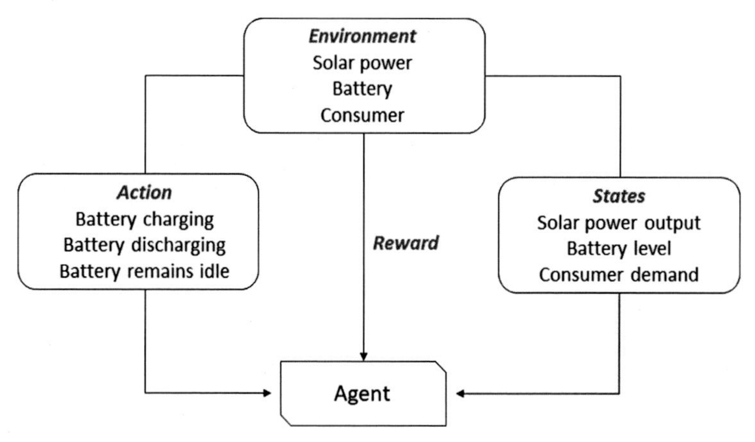

Fig. 11 RL Model.

2. Electrical energy consumption with time step t
3. Battery energy level with time step t

Actions of agent:

1. Battery charging
2. Battery discharging
3. Battery remains idle

Charge—discharge actions

1. Charging of battery is determines based on minimum charging rate, solar output and battery charging capacity
2. Battery discharging is based on minimum discharging rate and battery discharging capacity

Reward function helps to minimize the grid dependency for residential loads. This can be done by optimizing the battery. The Q value can be updated by following

- Initialize the Q value matrix, Q(s, a).
- Determine the current state, s
- Choose an action, a
- Observe the reward "r" and new state "s"
- Update the Q value i.e. $Q(s, a) = Q(s, a) + \alpha[r(s, a) + \gamma \max Q(s', a') - Q(s, a)]$
- Update the new state and repeat until terminal state is reached

7. Results and discussions

Based on three ANN models the model accuracy is determined and it is graphically presented Figs. 12–14 considering the RMSE.

In addition to these results the three model results are plotted based on real values (Fig. 15). It is inferred based on the obtained results that model 1

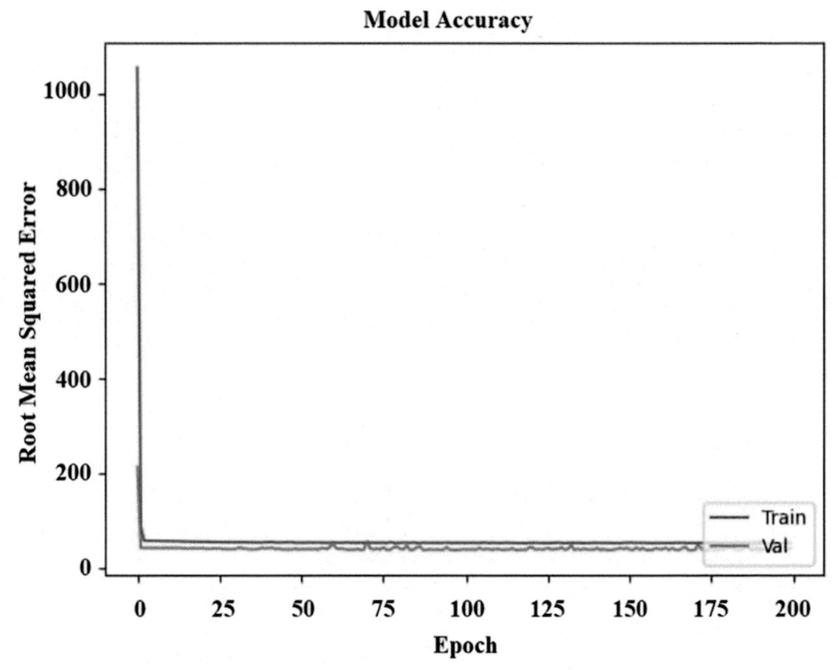

Fig. 12 Graphical analysis of Model 1 RMSE.

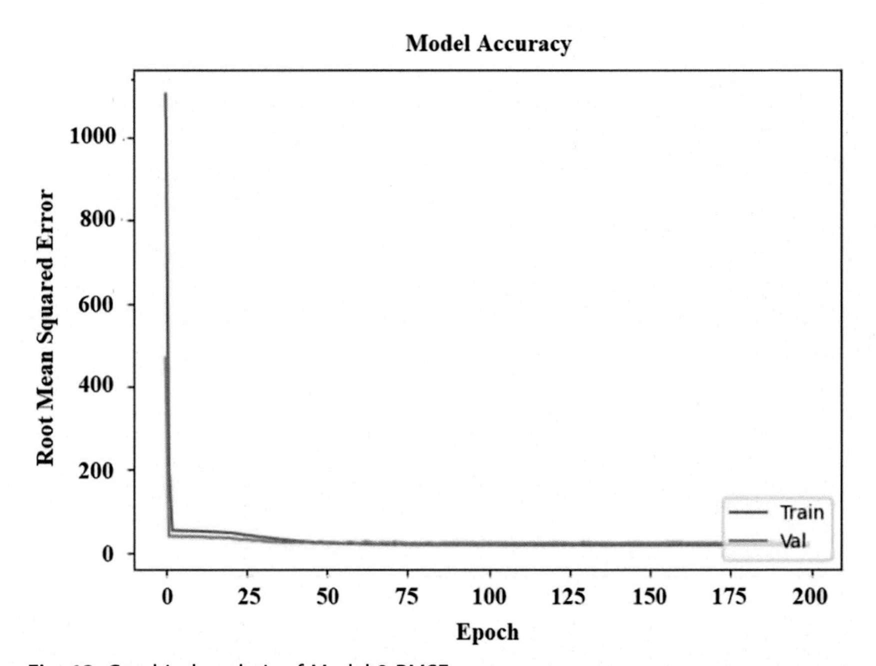

Fig. 13 Graphical analysis of Model 2 RMSE.

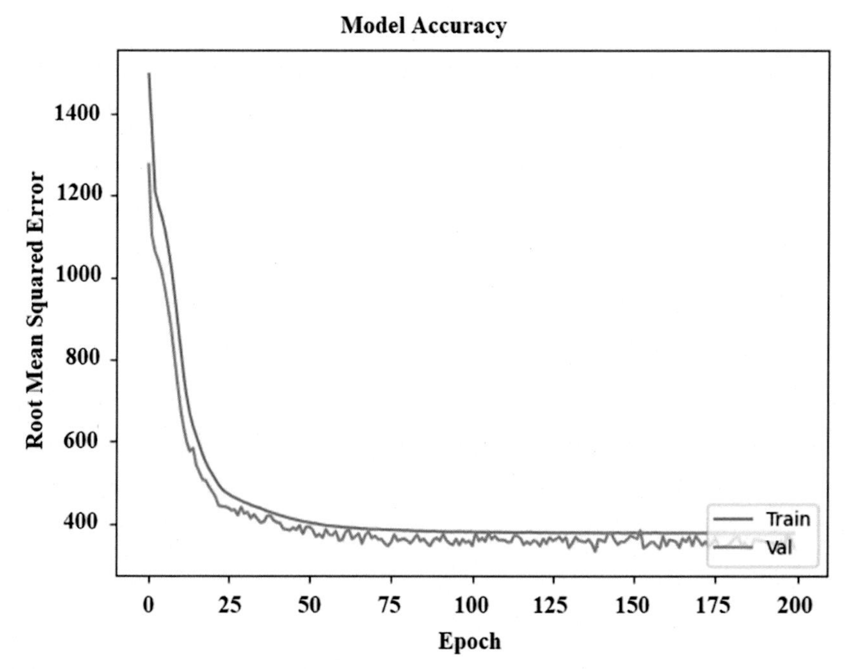

Fig. 14 Graphical analysis of Model 3 RMSE.

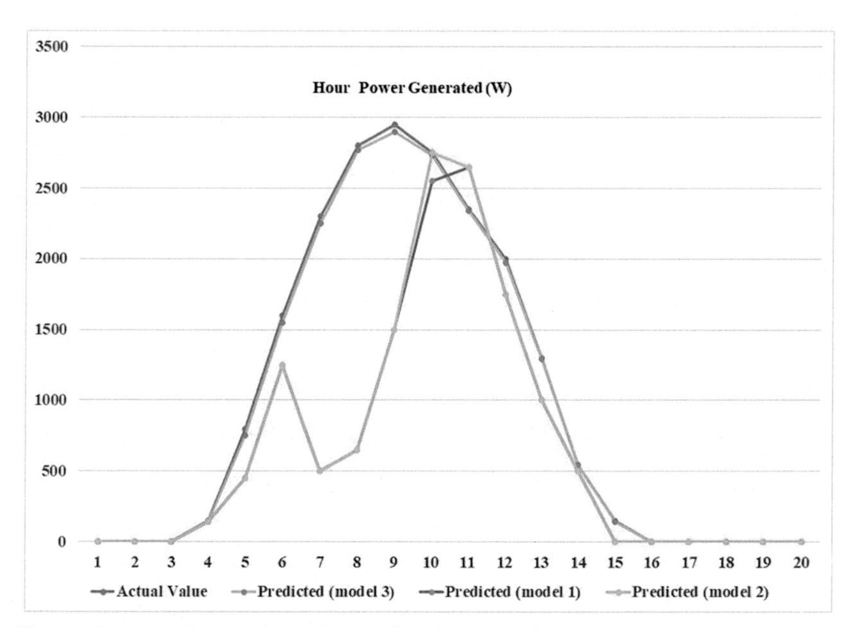

Fig. 15 Comparative results with actual and proposed system.

and 2 are predicted to get higher accuracy of 95% compared to actual value. In case model 3 it shows significant variation compared to actual value. Hence model 1 is used in our proposed system.

8. Conclusion

In this research work, a smart Energy Management System for residential load using Machine Learning techniques is discussed. Battery optimization strategies incorporated with ANN helps in predicting solar power generation and also in shifting loads between the grid and battery. The solar power generation is predicted by ANN and supervised learning technique. Battery usage optimization is implemented by Reinforcement Learning techniques. It is concluded from the obtained results that proposed REMS effectively the priority load list between grid and solar. In this proposed system the grid dependency is reduced at domestic building which results in reduction of CO_2 emission.

References

[1] A. Chatterjee, S. Paul, B. Ganguly, Multi-objective energy management of a smart home in real time environment, IEEE Trans. Indus. Appl. 50 (2023).
[2] B.K. Sethi, S.R. Amit Singh, D.S. Mohanty, R.K. Misra, Game theoretic smart residential buildings energy management system under false data injection attack, IEEE Int. Things J. 10 (2023).
[3] L. Renzhi, Z. Jiang, W. Huaming, Y. Ding, D. Wang, H.-T. Zhang, Reward shaping-based actor–critic deep reinforcement learning for residential energy management, IEEE Trans. Indus. Inform. 19 (2023).
[4] M. Prathapa Raju, A. Jaya Laxmi, Improved load management algorithm for EMU/HEMS using machine learning algorithms, Int. J. Elect. Eng. Technol. (IJEET) 9 (2018).
[5] P. Prakash, K. Panduranga, Design and development of advanced smart energy management system integrated with IoT framework, J. Energy Storage 25 (2019).
[6] N.K. Prakash, D. Vadana Prasanna, Machine learning based residential energy management system, IEEE Int. Conf. Comput. Intel. Comput. Res. (2017).
[7] S. Seyedzadeh, P.R. Farzad, G. Ivan, R. Marc, Machine learning for estimation of building energy consumption and performance: a review, Vis. Eng. 5 (2018).
[8] Solar Research, 2020. nrel.gov.
[9] S. Thomas, Deep Reinforcement Learning Course, 2020. simoninithomas.github.io.
[10] Machine Learning, and Deep Learning, Introduction to Tensor Flow for Artificial Intelligence, Machine Learning, and Deep Learning, Coursera, 2020.
[11] K. Team, Keras Documentation: Keras API Reference, Keras.io, 2020.
[12] Machine Learning, Coursera, 2020.
[13] Fundamentals of Reinforcement Learning, Coursera, 2020.
[14] Neural Networks and Deep Learning, Coursera, 2020.
[15] Solar Energy Basics, Coursera, 2020.

About the authors

Dr. R. Ramaprabha, Associate Professor in the Department of Electrical and Electronics Engineering, SSN College of Engineering, Chennai, has 23 years of teaching experience including 17 years of research experience. She obtained her PhD in the area of Solar Photovoltaic Systems from Anna University, Chennai. During her PhD, she developed "Solar Energy Research Lab," which is in the east wing of EEE Department (at SSN College of Engineering) with internal funding. She has published 115 research publications in international journals, 155 papers in the proceedings of refereed international and national conferences, 2 books, and 10 book chapters. She received the best teacher award for the academic years 2009–10, 2010–11, 2014–15, 2016–17, and 2020–21. She also received CTS best faculty award for the academic year 2016–17. She received "IET CLN Sir C. V. Raman Research Award 2014" in IET Chennai Network Achievements Awards 2014 function by IET. She guided six PhD students and guiding six PhD students. She is a reviewer for several referred international journals. She completed an internally funded project worth about ₹13 lakhs. She is a scientist mentor for the project titled "Design and development of flywheel-based power conditioning system for a renewable energy fed micro grid" of ₹21 lakhs funded by Department of Science and Technology (DST) under Women Scientists Scheme (WOS-A). She is a review board member for many reputed international journals. She holds the responsibility of IEEE Student Branch Counselor and NIRF Zonal Coordinator at institute level. She is an active senior member in IEEE and life member in ISTE.

Dr. G. Ramya, Assistant Professor in the Department of Computing Technologies, SRM Institute of Engineering and Technology, Chennai has 7 years of teaching experience including 6 years of research experience. She obtained her PhD in the area of Solar Photovoltaic System from Anna University, Chennai. She has published 19 research publications in international journals and conferences. She is a reviewer for several referred international journals.

CHAPTER FOUR

Text-based personality prediction using XLNet

Ashok Kumar Jayaraman[a], Gayathri Ananthakrishnan[b], Tina Esther Trueman[c], and Erik Cambria[d]

[a]Department of Information Science and Technology, Anna University, Chennai, India
[b]Department of Information Technology, VIT University, Vellore, India
[c]Department of Computer Science, University of the People, Pasadena, CA, United States
[d]School of Computer Science and Engineering, Nanyang Technological University, Singapore, Singapore

Contents

Abstract

Personality is a dynamic and organized set of characteristics that distinguish a person in thinking patterns, behaviors, emotions, and motivations based on biological and environmental factors. In particular, personality is a broad subject that is widely studied in various domains such as mental healthcare, web intelligence, and recommendation systems. Traditionally, researchers used psychology methods via psycholinguistic approaches (word counting in specific texts) to identify personality. Recently, social media data have been used for studies on personality. Psychologists and scientists are determining the personality of a person with the Myers–Briggs Type Indicator (MBTI) and Big Five model. Moreover, transformers-based models have shown better results in natural language processing tasks with context-dependent features. In this chapter, we propose a text-based personality prediction system using XLNet, which learns bidirectional context via factorization order and relative positional encoding. Experiments on two different gold-standard personality detection datasets show that the proposed model obtains up to 4% accuracy improvement.

Advances in Computers, Volume 132
ISSN 0065-2458
https://doi.org/10.1016/bs.adcom.2023.08.002

1. Introduction

Human personality is a colorful and complex system. It defines a person's feelings, behavior, and thinking patterns. In particular, personality is derived from an individual's experience and environmental factors. Therefore, the personality of an individual can change over time. However, adults relatively maintain their core personality traits during adulthood [1–3]. Earlier researchers used countless characteristics of an individual to determine personality. Nowadays, psychologists and scientists are determining the personality of a person with the Myers–Briggs Type Indicator (MBTI) and Big Five model. The MBTI is a personality type identification system that divides people into 16 distinct personality types across four axes, namely, Introversion (I)—Extroversion (E), Intuition (N)—Sensing (S), Thinking (T)—Feeling (F), and Judging (J)—Perceiving (P). First, I—E measures the outer vs inner world preference of an individual. Second, N—S measures the sensing vs impression patterns of an individual. Third, T—F determines objective principles and facts vs the emotional weights of an individual. Finally, J—P measures a planned and ordered life vs a spontaneous and flexible life of an individual.

The Big Five model is studied with five different personality types, namely, emotionality (or neuroticism), extroversion, agreeableness, conscientiousness, and openness (EXACO) [4, 5]. Neuroticism defines whether a person is sensitive, nervous, depressed, anxious, has negative feelings, and has self-doubt. Extroversion describes a person's talkativeness, outgoingness, and high energy. Agreeableness determines whether a person is cooperative, kind, trustworthy, polite, friendly, generous, and straightforward. Conscientiousness reflects whether a person is responsible, goal-directed, hard-working, and adheres strictly to norms and rules. Openness defines whether a person is curious about new ideas and new experiences. Some researchers use a six-factor model (HEXACO) for identifying personality. The six-factor model includes a new factor in addition to the Big Five model called honesty–humility. This personality trait reflects whether a person is moral, fair, sincere, and avoids greed [6].

Traditionally, researchers used psychology methods via psycholinguistic approaches (word counting in specific texts) to identify personality [7]. Psychology methods are broadly studied in two categories, namely, qualitative and quantitative. These categories are used in the form of case study, experiment, observational study, survey, and content analysis. In recent years, social media has become popular among internet users. They express their feelings and views in the form of audio, video, image, and text. These data become complex in nature to identify personality. Therefore, researchers used natural language processing (NLP), conventional machine learning, and deep learning methods to identify personality. They learn BoW (Bag of words) features and semantic context features. Moreover, the recurrent neural networks in deep learning capture a unidirectional context, i.e., from beginning to end and from end to beginning. However, Vaswani et al. [8] and Devlin et al. [9] introduced attention-based models such as transformer (encoder–decoder structure) and BERT (encoder structure) to capture bidirectional context with sequential parallelization. Later, Yang et al. [10] introduced a permutation language model (also called XLNet) to capture bidirectional context via factorization order and relative positional encoding. It has shown promising results over the BERT pretrained model. Thus, we propose a permutation language modeling for personality detection. This chapters contributes to the following.

- Addresses the MBTI and Big Five model for personality detection
- Employs a permutation language pretrained model
- Outperforms the personality detection task than the state-of-the-art models

The rest of this chapter is organized as follows: Section 2 describes the related works; Section 3 presents the text-based personality detection framework based on XLNet; Section 4 presents results and discussion; and finally, Section 5 offers concluding remarks.

2. Related works

Researchers studied the automated personality type detection system using various machine learning and deep learning models. The personality type detection system is used in various applications such as managerial style prediction [11], academic major selection [12], dentists' career choice prediction [13], college performance prediction system [14]. In particular, we discuss the text-based research works on MBTI and the Big Five model. Asghar et al. [15] investigated an attention-based BiLSTM for psychopath personality trait detection using social media text such as Facebook and Twitter. Their study indicated that the attention-based Bi-LSTM model achieves 85% accuracy for detecting psychopath and nonpsychopath. Stachl et al. [16] examined the Big Five personality model with six different behavioral information of an individual. This behavioral information is collected from smartphones via sensor and log data. Their study indicated that there are specific behavioral patterns in app usage, music consumption, mobility, day and night activity, and overall phone activity. Also, the authors suggested that there are benefits in terms of research and danger in terms of privacy implications. Amirhosseini and Kazemian [17] developed the MBTI-based automated personality type prediction system using the XGBoost algorithm. Specifically, the authors divided the 16 distinct classes into 4 binary classification tasks. They trained the binary classifier using TF-IDF features on each of the personality types separately. Their study indicated that the XGBoost model has shown reliability and better accuracy. Hernandez and Scott [18] performed the recurrent neural network to classify the MBTI personality type dataset. The authors divided the 16 MBTI classes into 4 binary tasks. Therefore, they trained four binary classifiers instead of a multiclass classifier. Their results indicated that the proposed RNN model achieves an average accuracy of 54.4% for the posts classification and 67.8% for user classification.

Ozer and Benet-Martinez [19] studied the personality characteristics and consequential outcomes using various meta-analyses. Their study claimed that the personality effects are influencing each of us while aggregating at the population level. Li et al. [20] proposed a multitask learning framework to predict emotional behavior and personality traits. Their study indicated that the convolutional neural network (CNN) achieves better performance on various personality and emotion datasets. Kazameini et al. [21] developed a bagged SVM model using contextualized word embedding for personality trait detection. The authors have broken the essay dataset into multiple chunks to extract maximum information. These subdocuments are associated with the same personality type. Later, they performed an SVM binary classifier to predict the corresponding personality trait. This method achieves 59.03% accuracy on average. Poria et al. [22] proposed a common-sense knowledge-based architecture to detect personality from the text. They achieved 63.6% average accuracy using the support vector machine (SVM). Majumder et al. [23] presented a CNN-based personality trait detection system. Their study indicated that CNN outperforms for all five personality traits with different configurations, and n-gram models showed no improvement. In summary, the existing researchers studied the personality type prediction system using a bag of word features, context-independent features, and context-dependent features. In this chapter, we propose the text-based personality detection system using permutation language modeling, where it supports context-dependent features via factorization order.

3. Method

In this section, we present the text-based personality prediction system using XLNet as shown in Fig. 1. We explain each part of this model as follows.

3.1 Datasets

We use two datasets for the personality identification system such as the MBTI dataset [24] and Essays dataset [25]. First, we use the MBTI dataset for the personality type identification system. This dataset groups each individual into 16 distinct personalities across four axes, namely, Introversion (I)—Extroversion (E), Intuition (N)—Sensing (S), Thinking (T)—Feeling (F),

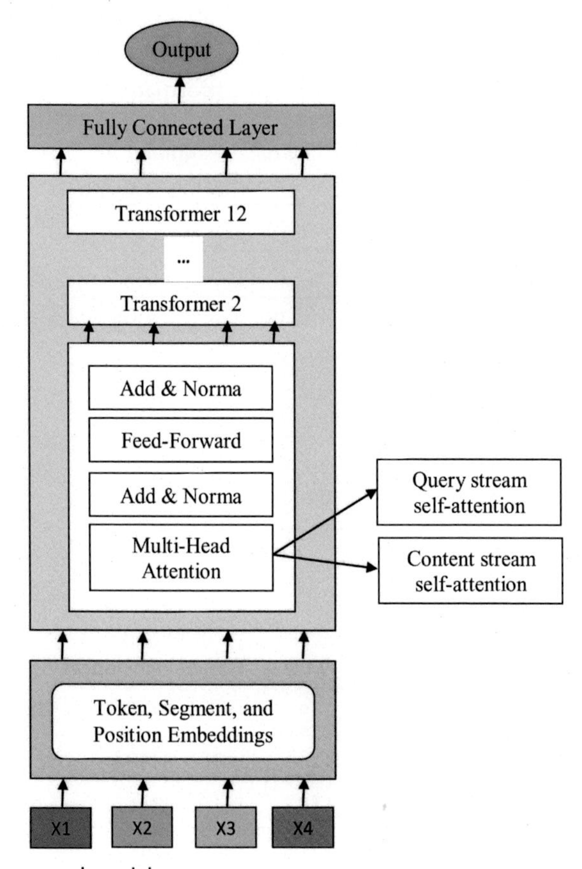

Fig. 1 The proposed model.

and Judging (J)—Perceiving (P). In particular, the MBTI dataset contains 8675 user posts associated with their personality type as in Table 1. Each personality type is encoded with four letters like INTJ (introversion, intuition, thinking, and judging). Second, we use the Essays dataset for the personality type identification system. This dataset contains 2467 anonymous essays associated with the Big Five personality measures such as emotionality (or neuroticism), extroversion, agreeableness, conscientiousness, and openness (EXACO) as in Table 2.

3.2 XLNet

XLNet is one of the latest pretrained models that achieve promising results in various NLP tasks such as document ranking, question answering, and sentiment analysis. Let $x = [x_1, x_2, \ldots, x_T]$ be the given input text with length T. Then, autoregressive language model ARLM maximizes the likelihood in a unidirectional way, i.e., from the forward or backward factorization order. This model fails to capture context information from both forward

Table 1 MBTI personality data distribution.

Personality	#instances	Personality	#instances
ENFJ	190	INFJ	1470
ENFP	675	INFP	1832
ENTJ	231	INTJ	1091
ENTP	685	INTP	1304
ESFJ	42	ISFJ	166
ESFP	48	ISFP	271
ESTJ	39	ISTJ	205
ESTP	89	ISTP	337

Table 2 Essays personality data distribution.

Personality	N Class	Y Class	#instances
Extroversion	1191	1276	2467
Neuroticism	1234	1233	2467
Agreeableness	1157	1310	2467
Conscientiousness	1214	1253	2467
Openness	1196	1271	2467

and backward directions. Therefore, ELMo [26] is introduced to capture context information from both forward and backward directions using a bidirectional language model. However, this model concatenates the contextual information from both directions, which are separately learned from the forward and backward directions. On the other hand, autoencoding language model learns contextual information from both forward and backward directions. Specifically, the BERT language model is designed to learn context from its surrounding information. This model predicts the masked words or tokens in the given input text. However, the BERT model has some critical aspects such as independence assumption between masked tokens, input noise, and context dependency. To address these critical aspects, Yang et al. [10] proposed a generalized autoregressive language model (XLNet). The XLNet uses factorization order and positional encoding to learn bidirectional context information. The objective function of this model is mathematically defined as in Eq. (1).

$$\max_{\theta} \; E_{z \sim Z_T}\left[\sum_{t=1}^{T} \log p_\theta\left(x_{z_t} | X_{z_{<t}}\right)\right] \tag{1}$$

where Z_T denotes the set of all possible permutations of the input sequence with length T. z_t denotes the current $(t - th)$ element. $z_{<t}$ denotes the previous elements $(t - 1)$ of a permutation $z \in Z_T$. The objective function takes the previous elements as the input context for predicting the current element or token. In particular, the objective function considers the permutation-based factorization order rather than the input sequence order. Let $S =$ *I*, *like*, *this*, *news* be the input text. Then, the permutation order of this input text is 4! Let [3 → 1 → 2 → 4] and [1 → 2 → 4 → 3] be the two sequence orders. Now, we predict the target element "this," which is computed as *P*(*this*). This target element appears first in the first sequence order and last in the second sequence order. However, the first sequence order has no preceded elements to look over and the second sequence order looks at all preceded elements to compute the probability of target element as *P*(*this* | *I*, *like*, *news*). Moreover, the transformer-based XLNet architecture uses two-stream self-attention mechanisms such as content stream and query stream for achieving this kind of permutation language modeling. First, the content stream self-attention mechanism encodes the context and content in the form of the standard hidden state representations as in Eq. (2). Second, the query stream self-attention mechanism encodes the context and positional information as in Eq. (3).

$$h_{z_t}^{(m)} \leftarrow Attention(Q = h_{z_t}^{(m-1)}, KV = h_{z_{\leq t}}^{(m-1)}; \theta) \tag{2}$$

$$g_{z_t}^{(m)} \leftarrow Attention(Q = g_{z_t}^{(m-1)}, KV = h_{z_{<t}}^{(m-1)}; \theta) \tag{3}$$

where Q, K, and V denotes the query, key, and value, and m denotes the number of attention layers.

In addition, XLNet reduces the optimization problem in the autoregressive language model using partial prediction. This predicts the last elements in a factorization order. Therefore, the log-likelihood of a target subsequence ($z_{>c}$) is maximized that conditioned on a nontarget subsequence ($z_{\leq c}$) as in Eq. (4). Here, c is the cutting point of a subsequence. ($Z_{>c}$) possesses the longest contextual information in the factorization order (z). Moreover, XLNet adopts the relative positional encoding and segment recurrence mechanism from Transformer-XL. These two techniques help XLNet to improve the long-term dependency of the given input text. Therefore, XLNet pretrained model is used as an effective model for NLP tasks.

$$\max_{\theta} \; E_{z \sim Z_T}\left[\log p_\theta(x_{z_{>c}} | X_{z_{\leq c}})\right] = E_{z \sim Z_T}\left[\sum_{t=c+1}^{|z|} \log p_\theta(x_{z_t} | X_{z_{<t}})\right] \tag{4}$$

In addition, XLNet incorporates two important techniques from Transformer-XL, namely, relative positional encoding and segment recurrence mechanism. These techniques help to improve the long-range dependency of the input sequence. Therefore, XLNet is used as an effective pretrained language model for NLP tasks.

3.3 XLNet fine-tuning

We use the XLNet pretrained model for identifying the personality of an individual from text. It is developed in two variants, namely, XLNet base model and XLNet large model. The XLNet base model represents 110M parameters with 12 transformer layers, 12 attention heads, and 768 hidden state units, and the XLNet large model represents 340M parameters with 24 transformer layers, 24 attention heads, and 1024 hidden state units. These architectures represent the same as BERT base and BERT large model. In particular, the BERT is build with the encoder structure of the transformer, and XLNet is build with the decoder structure of the transformer. The XLNet accepts [*CLS, A, SEP, B, SEP*] as the input format, where [*CLS*] and [*SEP*] are special elements or tokens to represent a classification token and sentence differentiation, and [*A*] and [*B*] are input segments. In the personality detection task, the given input sequence is aggregated by the token embedding, relative segment embedding, and relative position embedding. This aggregated input representation is fed to the transformer blocks. Then, an output layer with softmax activation function is added on the top of the XLNet transformers for calculating the output probabilities for each category.

4. Results and discussion

We experimented with the XLNet base model in Google Colab Pro on two different personality detection datasets, namely, the MBTI dataset and essays dataset. The MBTI dataset contains 8675 user posts, where each post is assigned with 1 of the 16 distinct personality types. We used the stratified random sampling method to divide the dataset into training (7026), validation (781), and testing (868) as in Table 3. Similarly, the essays dataset

Table 3 MBTI personality data split for training, validation, and testing.

Personality	Training	Validation	Testing
ENFJ	154	17	19
ENFP	546	61	68
ENTJ	187	21	23
ENTP	554	62	69
ESFJ	34	4	4
ESFP	39	4	5
ESTJ	31	4	4
ESTP	72	8	9
INFJ	1191	132	147
INFP	1484	165	183
INTJ	884	98	109
INTP	1057	117	130
ISFJ	134	15	17
ISFP	220	24	27
ISTJ	166	19	20
ISTP	273	30	34
Total	7026	781	868

Table 4 Essays personality data split for training, validation, and testing.

	Train (#1998)		Valid (#222)		Test (#247)		
Personality	**N**	**Y**	**N**	**Y**	**N**	**Y**	**Total**
Extroversion	965	1033	107	115	119	128	2467
Neuroticism	999	999	111	111	124	123	2467
Agreeableness	937	1061	104	118	116	131	2467
Conscientiousness	983	1015	109	113	122	125	2467
Openness	968	1030	108	114	120	127	2467

contains 2467 essays, where each essay is associated with the Big-Five personality type. This dataset is divided into training (1998), validation (222), and testing (247) for each personality type using the stratified random sampling as in Table 4. For instance, the extroversion personality type is divided into 80:10:10 for training (1998), validation (222), and testing (247). Specifically, we expanded the shorten texts to full text (e.g., "aren't" into "are not") using contraction map dictionary and removed punctuations except for the period, question mark, and exclamation mark. We then represent these datasets into the input tokens, segment tokens, and position tokens. These tokens are aggregated and fed into the proposed XLNet base pretrained model. In particular, we conduct a 16-class classification task on the MBTI dataset and five 2-class classification tasks on the essays dataset. For fine-tuning task, we used Adam optimizer with learning rate 2e-5, 1000 input sequence lengths, 8 epochs, 4 batch sizes, and 14,056 training steps for the MBTI dataset, and Adam optimizer with learning rate 2e-6, 1000 input sequence lengths, 8 epochs, 4 batch sizes, and 4000 training steps for the essays dataset.

Moreover, the XLNet base model is constructed with 110M parameters. It learns bidirectional context information via relative positional encoding and factorization order. We use the standard evaluation metrics such as precision (P), recall (R), and F1-score (F1) and their micro average, macro average, and weighted average [29] for computing the performance of the personality detection task. Table 5 shows the obtained validation and testing results for the MBTI dataset. This table indicates that the proposed fine-tuning model achieves 72% for micro precision, micro recall, and micro F1-score for personality detection tasks in the validation and testing datasets. Tables 6 and 7 show the five 2-class personality detection performances for validation and testing. In these tables, the openness personality type achieves 64% for micro precision, micro recall, and micro F1-score for both validation and testing datasets. It is higher than the other personality types.

Table 5 Performance of the MBTI dataset.

	Valid			Test		
Personality	**P**	**R**	**F1**	**P**	**R**	**F1**
ENFJ	0.82	0.82	0.82	0.65	0.79	0.71
ENFP	0.78	0.74	0.76	0.72	0.75	0.73
ENTJ	0.74	0.67	0.70	0.67	0.78	0.72
ENTP	0.64	0.68	0.66	0.76	0.70	0.73
ESFJ	0.67	1.00	0.80	1.00	0.75	0.86
ESFP	0.50	0.25	0.33	0.67	0.40	0.50
ESTJ	0.00	0.00	0.00	0.80	1.00	0.89
ESTP	1.00	0.50	0.67	0.80	0.44	0.57
INFJ	0.76	0.66	0.70	0.79	0.70	0.74
INFP	0.70	0.84	0.77	0.68	0.83	0.75
INTJ	0.69	0.73	0.71	0.76	0.65	0.70
INTP	0.68	0.70	0.69	0.71	0.75	0.73
ISFJ	0.71	0.67	0.69	0.74	0.82	0.78
ISFP	0.73	0.67	0.70	0.57	0.48	0.52
ISTJ	0.87	0.68	0.76	0.71	0.50	0.59
ISTP	0.89	0.57	0.69	0.76	0.65	0.70
Micro	0.72	0.72	0.72	0.72	0.72	0.72
Macro	0.70	0.64	0.65	0.74	0.69	0.70
Weighted	0.72	0.72	0.71	0.73	0.72	0.72

The result comparison for the MBTI dataset is shown in Table 8. In this table, Varma [27] performed various machine learning models in the ratio of 60:40 (training: testing) and 70:30. The authors indicated that the logistic regression model achieves 58.19% accuracy in the 60:40 ratio, and the logistic regression and XG Boost models achieve 57.12% accuracy in the 70:40 ratio. Hernandez et al. [18] performed LSTM network as five 2-class classifications on the MBTI dataset and achieved 54.40% accuracy. Uzsoy [30] performed the BERT pretrained model using TPU (Tensor Processing Unit). The authors achieved 68.10% accuracy with 1500 input sequence lengths. Our proposed XLNet base model achieves 72% accuracy using GPU. Then, the result comparison for the Big Five personality dataset

Table 6 Validation performance of the essays dataset.

Personality	Extroversion			Neuroticism			Agreeableness			Conscientiousness			Openness		
Class	P	R	F1	P	R	F1	P	R	F1	P	R	F1	P	R	F1
N	0.57	0.61	0.59	0.63	0.50	0.55	0.70	0.18	0.29	0.62	0.49	0.54	0.64	0.60	0.62
Y	0.61	0.57	0.59	0.58	0.70	0.64	0.56	0.93	0.70	0.59	0.71	0.64	0.64	0.68	0.66
Micro	0.59	0.59	0.59	0.60	0.60	0.60	0.58	0.58	0.58	0.60	0.60	0.60	0.64	0.64	0.64
Macro	0.59	0.59	0.59	0.60	0.60	0.59	0.63	0.56	0.50	0.60	0.60	0.59	0.64	0.64	0.64
Weighted	0.59	0.59	0.59	0.60	0.60	0.59	0.63	0.58	0.51	0.60	0.60	0.59	0.64	0.64	0.64

Table 7 Testing performance of the essays dataset.

Personality	Extroversion			Neuroticism			Agreeableness			Conscientiousness			Openness		
Class	P	R	F1	P	R	F1	P	R	F1	P	R	F1	P	R	F1
N	0.51	0.66	0.61	0.60	0.54	0.57	0.69	0.21	0.32	0.63	0.52	0.57	0.62	0.64	0.63
Y	0.62	0.53	0.57	0.58	0.64	0.61	0.57	0.92	0.70	0.60	0.70	0.64	0.65	0.63	0.64
Micro	0.59	0.59	0.59	0.59	0.59	0.59	0.58	0.58	0.58	0.61	0.61	0.61	0.64	0.64	0.64
Macro	0.59	0.59	0.59	0.59	0.59	0.59	0.63	0.56	0.51	0.61	0.61	0.61	0.64	0.64	0.64
Weighted	0.60	0.59	0.59	0.59	0.59	0.59	0.62	0.58	0.52	0.61	0.61	0.61	0.64	0.64	0.64

Table 8 Result comparison for the MBTI dataset.

Authors	Model	Test
Varma (60:40) [27]	RandomForest	39.53
	XG Boost	57.87
	Gradient Descent	41.10
	Logistic Regression	58.19
	KNN	16.45
	SVM	35.52
Varma (70-30) [27]	RandomForest	37.87
	XG Boost	57.12
	Gradient Descent	45.13
	Logistic Regression	57.12
	KNN	16.96
	SVM	35.76
Hernandez et al. [18]	LSTM	54.40
Uzsoy [30]	BERT	68.10
Proposed	XLNet-Base-FT	72.24

Table 9 Result comparison for the essays dataset.

Personality	Majority baseline [21, 23]	Mairesse [23, 28]	CNN + Mairesse [21, 23]	BB-SVM [21]	Proposed	
					Valid	Test
Extroversion	51.72	55.13	58.09	59.30	59.01	59.10
Neuroticism	50.20	58.90	57.33	59.39	59.91	59.11
Agreeableness	53.10	55.35	56.71	56.52	58.11	58.30
Conscientiousness	50.79	55.28	56.71	57.84	59.91	61.13
Openness	51.52	59.57	61.13	62.09	64.41	63.56
Average	51.43	56.84	57.99	59.03	**60.27**	**60.24**

is shown in Table 9. Our proposed method outperforms than the majority baseline [21, 23], Mairesse [23, 28], CNN and Mairesse [21, 23], and BB-SVM [21] models. In the BB-SVM model, the authors had broken the essays into multiple subdocuments with 200 input sequence tokens. Therefore, we indicate that our proposed XLNet fine-tuning model achieves better results in both datasets.

5. Conclusion

In this chapter, we presented a text-based personality prediction system using permutation language modeling. This model captures bidirectional context-dependent features via factorization order and positional encoding. In particular, we expanded the shorten texts and removed punctuations except for the period, exclamation mark, and question mark. These help us to maintain the segment context. We then performed a 16-class personality detection system on the MBTI dataset and five 2-class personality detection systems on the essays dataset. We evaluated the proposed fine-tuning model using precision, recall, F1-score, and their macro average, micro average, and weighted average scores. Our results indicate that the proposed model achieves a 72% micro F1-score on the MBTI dataset and a 60% micro F1-score on the essays dataset. In future work, we desire to study the personality of an individual in cybercrime activity in a large dataset with gender and age group.

Acknowledgments

This work was supported by the University Grants Commission (UGC), Government of India under the National Doctoral Fellowship.

References

[1] Psychology Today website, Personality and its traits, https://www.psychologytoday.com/us/basics/personality, (accessed 05.06.21).
[2] M. Gjurković, J. Šnajder, Reddit: a gold mine for personality prediction, in: Proceedings of the Second Workshop on Computational Modeling of People's Opinions, Personality, and Emotions in Social Media, 2018, pp. 87–97.
[3] A.H. Buss, Personality as traits, Am. Psychol. 44 (11) (1989) 1378.
[4] J.M. Digman, Personality structure: emergence of the five-factor model, Annu. Rev. Psychol. 41 (1) (1990) 417–440.
[5] Y. Mehta, N. Majumder, A. Gelbukh, E. Cambria, Recent trends in deep learning based personality detection, Artif. Intell. Rev. 53 (2020) 2313–2339.
[6] M.C. Ashton, K. Lee, M. Perugini, P. Szarota, R.E. De Vries, L. Di Blas, K. Boies, B. De Raad, A six-factor structure of personality-descriptive adjectives: solutions from psycholexical studies in seven languages, J. Pers. Soc. Psychol. 86 (2) (2004) 356.
[7] S. Štajner, S. Yenikent, A survey of automatic personality detection from texts, in: Proceedings of the 28th International Conference on Computational Linguistics, 2020, pp. 6284–6295.
[8] A. Vaswani, N. Shazeer, N. Parmar, J. Uszkoreit, L. Jones, A.N. Gomez, L. Kaiser, I. Polosukhin, Attention is all you need, arXiv:1706.03762 (2017).
[9] J. Devlin, M.-W. Chang, K. Lee, K. Toutanova, Bert: pre-training of deep bidirectional transformers for language understanding, arXiv:1810.04805 (2018).
[10] Z. Yang, Z. Dai, Y. Yang, J. Carbonell, R. Salakhutdinov, Q.V. Le, Xlnet: Generalized autoregressive pretraining for language understanding, arXiv:1906.08237 (2019).

[11] E.Y.D. Sari, K. Bashori, Predicting managerial styles: is the Myers-Briggs type indicator still useful? J. Educ. Learn. (EduLearn) 14 (4) (2020) 617–622.
[12] C.A. Pulver, K.R. Kelly, Incremental validity of the Myers-Briggs type indicator in predicting academic major selection of undecided university students, J. Career Assess. 16 (4) (2008) 441–455.
[13] T.G. Grandy, G.H. Westerman, R.A. Ocanto, C.G. Erskine, Predicting dentists'career choices using the Myers-Briggs type indicator, J. Am. Dent. Assoc. 127 (2) (1996) 253–258.
[14] L.J. Stricker, H. Schiffman, J. Ross, Prediction of college performance with the Myers-Briggs type indicator, Educ. Psychol. Meas. 25 (4) (1965) 1081–1095.
[15] J. Asghar, S. Akbar, M.Z. Asghar, B. Ahmad, M.S. Al-Rakhami, A. Gumaei, Detection and classification of psychopathic personality trait from social media text using deep learning model, Comput. Math. Methods Med. 2021 (2021) 1–10.
[16] C. Stachl, Q. Au, R. Schoedel, S.D. Gosling, G.M. Harari, D. Buschek, S.T. Völkel, T. Schuwerk, M. Oldemeier, T. Ullmann, Predicting personality from patterns of behavior collected with smartphones, Proc. Natl. Acad. Sci. U.S.A 117 (30) (2020) 17680–17687.
[17] M.H. Amirhosseini, H. Kazemian, Machine learning approach to personality type prediction based on the Myers-Briggs type indicator®, Multimodal Technol. Interact. 4 (1) (2020) 9.
[18] R.K. Hernandez, I. Scott, Predicting Myers-Briggs type indicator with text, in: 31st Conference on Neural Information Processing Systems (NIPS 2017), 2017.
[19] D.J. Ozer, V. Benet-Martinez, Personality and the prediction of consequential outcomes, Annu. Rev. Psychol. 57 (2006) 401–421.
[20] Y. Li, A. Kazameini, Y. Mehta, E. Cambria, Multitask Learning for Emotion and Personality Detection, arXiv:2101.02346 (2021).
[21] A. Kazameini, S. Fatehi, Y. Mehta, S. Eetemadi, E. Cambria, Personality trait detection using bagged SVM over BERT word embedding ensembles, in: Proceedings of WiNLP, 2020.
[22] S. Poria, A. Gelbukh, B. Agarwal, E. Cambria, N. Howard, Common sense knowledge based personality recognition from text, in: Mexican International Conference on Artificial Intelligence, Springer, 2013, pp. 484–496.
[23] N. Majumder, S. Poria, A. Gelbukh, E. Cambria, Deep learning-based document modeling for personality detection from text, IEEE Intell. Syst. 32 (2) (2017) 74–79.
[24] MBTI Kaggle dataset, https://www.kaggle.com/datasnaek/mbti-type, (accessed 27.05.21).
[25] J.W. Pennebaker, L.A. King, Linguistic styles: language use as an individual difference, J. Pers. Soc. Psychol. 77 (6) (1999) 1296.
[26] M.E. Peters, M. Neumann, M. Iyyer, M. Gardner, C. Clark, K. Lee, L. Zettlemoyer, Deep contextualized word representations, arXiv:1802.05365 (2018).
[27] R. Varma, MBTI personality predictor using machine learning, 2021. https://www.kaggle.com/rajshreev/mbti-personality-predictor-using-machine-learning.
[28] F. Mairesse, M.A. Walker, M.R. Mehl, R.K. Moore, Using linguistic cues for the automatic recognition of personality in conversation and text, J. Artif. Intell. Res. 30 (2007) 457–500.
[29] J.A. Kumar, S. Abirami, A. Ghosh, T.E. Trueman, A C-LSTM with attention mechanism for question categorization, in: Symposium on Machine Learning and Metaheuristics Algorithms, and Applications, Springer, 2019, pp. 234–244.
[30] A.S. Uzsoy, Myers-Briggs types with tensorflow/BERT, 2020. https://www.kaggle.com/anasofiauzsoy/myers-briggs-types-with-tensorflow-bert.

About the authors

Ashok Kumar Jayaraman is a research scholar with the Department of Information Science and Technology, Anna University, Chennai. He received a Bachelor of Science in Mathematics from the University of Madras and a Master of Computer Applications from Anna University. Also, He has received UGC National Fellowship and Indian National Science Academy (INSA) Visiting Scientist Research Fellowship. His research interests include natural language processing, sentiment analysis, machine learning, deep learning, and social network analysis.

Gayathri Ananthakrishnan is currently working as an Assistant Professor (Senior) in the School of Computer Science Engineering and Information Systems (SCORE), VIT, Vellore. She has obtained her degree at Anna University. She has participated and presented at several National, International Conferences and Seminars. Her areas of interest include Wireless Network Security, Machine Learning, the Internet of Things (IoT), and Sentiment Analysis.

Tina Esther Trueman is an Assistant Professor with the Department of Computer Science, University of People, California. She obtained her M.E. in Computer Science and Engineering, Anna University, and obtained Ph.D. in Cloud data analytics from Anna University. She has published research papers in refereed journals and conferences. Her research interests include fuzzy logic systems, text mining, machine learning, deep learning, data mining, and data analytics.

Erik Cambria is an Associate Professor at NTU, where he also holds the appointment of Provost Chair in Computer Science and Engineering. Prior to joining NTU, he worked at Microsoft Research Asia (Beijing) and HP Labs India (Bangalore) and earned his PhD through a joint program between the University of Stirling and MIT Media Lab. His research focuses on the ensemble application of symbolic and sub-symbolic AI to natural language processing tasks such as sentiment analysis, dialogue systems, and financial forecasting. Erik is recipient of many awards, e.g., the 2019 IEEE Outstanding Early Career Award, he was listed among the 2018 AI's 10 to Watch, and was featured in Forbes as one of the 5 People Building Our AI Future. He is Associate Editor of several journals, e.g., INFFUS, IEEE CIM, and KBS, Special Content Editor of FGCS, Department Editor of IEEE Intelligent Systems, and is involved in many international conferences as program chair and invited speaker.

Articulating the power of reasoning and gathering data for information security and justice

Preetha Evangeline David[a] and P. Anandhakumar[b]

[a]Department of Artificial Intelligence and Machine Learning, Chennai Institute of Technology, Chennai, Tamil Nadu, India

[b]Department of Information Technology, Madras Institute of Technology, Anna University, Chennai, Tamil Nadu, India

Contents

Abstract

With the accumulation of big data, dramatic improvements in computing power, and continuous innovation in Machine Learning (ML) methods, Artificial Intelligence (AI) technologies such as image recognition, voice recognition, and natural language processing have become ubiquitous. Meanwhile, AI poses a significant impact on

Advances in Computers, Volume 132
ISSN 0065-2458
https://doi.org/10.1016/bs.adcom.2023.08.003

computer security: on the one hand, AI can be used to build defensive systems such as malware and network attack detection; on the other hand, AI might be exploited to launch more effective attacks. In some scenarios, the security of AI systems is a matter of life and death. Thus, building robust AI systems that are immune to external interference is essential. AI can benefit security and vice versa. Unlike security vulnerabilities in traditional systems, the root cause of security weaknesses in ML systems lies in the lack of explicability in AI systems. This lack of explicability leaves openings that can be exploited by adversarial machine learning methods such as evasion, poisoning, and backdoor attacks. These attacks are very effective and have strong transferability among different ML models, and thus pose serious security threats to Deep Neural Network (DNN)-based AI applications. There is a long way to go before the industry achieves secure and robust AI systems. On the technology side, we need to continuously research AI explicability to understand how to implement a sound AI foundation and build systematic defense mechanisms for secure AI platforms.

1. Introduction

With the accumulation of big data, dramatic improvements in computing power and continuous innovation in Machine Learning (ML) methods, Artificial Intelligence (AI) technologies such as image recognition, voice recognition, and natural language processing have become ubiquitous. The development and extensive commercial use of AI technologies have heralded the coming of an intelligent world. Today, with development of chips and sensors, the intention to "Activate Intelligence" is affecting a broad range of sectors, including:

- Intelligent transportation—choose the best route
- Intelligent healthcare—understand your health
- Intelligent manufacturing—adapt to your needs

AI in the past two decades can be attributed to the following reasons [1]: (1) Massive amounts of data: With the rise of the Internet, data grows rapidly in the form of voice, video, and text. This data provides sufficient input for ML algorithms, prompting the rapid development of AI technologies. (2) Scalable computer and software systems: The success of deep learning in recent years is mainly driven by new dedicated hardware, such as CPU clusters, GPUs, and Tensor Processing Units (TPUs), as well as software platforms. (3) The broad accessibility of these technologies: A large amount of open-source software now bolsters data processing and AI- related work, reducing development time and costs. In addition, many cloud services provide developers with readily available computing and storage resources. In several types of deployments, such as robots, virtual assistants, autonomous driving, intelligent transportation, smart manufacturing, and Smart Cities, AI is moving toward historic accomplishments. Big companies such as Google, Microsoft, and Amazon have taken AI as their core strategy for future development. In 2017, Google DeepMind unveiled the AlphaGo Zero, which honed its Go-playing prowess simply by playing games against itself and defeated the champion-defeating version of AlphaGo after only three days of self-training. AlphaGo Zero is able to discover new knowledge and develop rule-breaking policies, revealing the tremendous potential of using AI technologies to change human life. What we see now is only a start. In the future, we will have a fully connected smart world. AI will bring superior experience to people everywhere, positively affect our work and life, and boost economic prosperity.

2. Five challenges to AI security

AI has great potential to build a better, smarter world, but at the same time faces severe security risks. Due to the lack of security consideration at the early development of AI algorithms, attackers are able to manipulate the inference results in ways that lead to misjudgment. In critical domains such as healthcare, transportation, and surveillance, security risks can be devastating. Successful attacks on AI systems can result in property loss or endanger personal safety. AI security risks exist not only in theoretical analyses but also in AI deployments. For instance, attackers can craft files to bypass AI-based detection tools or add noise to smart home voice control command to invoke malicious applications. Attackers can also tamper with data returned by a terminal or deliberately engage in malicious dialogs with a chat robot to cause a prediction error in the backend AI system. It is even possible to apply small stickers on traffic signs or vehicles that cause false inferences by autonomous vehicles.To mitigate these AI security risks, AI system design must overcome five security challenges:

- Software and hardware security: The code of applications, models, platforms, and chips may have vulnerabilities or backdoors that attackers can exploit. Further, attackers may implant backdoors in models to launch advanced attacks. Due to the inexplicability of AI models, the backdoors are difficult to discover.
- Data integrity: Attackers can inject malicious data in the training stage to affect the inference capability of AI models or add a small perturbation to input samples in the inference stage to change the inference result.

- Model confidentiality: Service providers generally want to provide only query services without exposing the training models. However, an attacker may create a clone model through a number of queries.
- Model robustness: Training samples typically do not cover all possible corner cases, resulting in the insufficiency of robustness. Therefore the model may fail to provide correct inference on adversarial examples.
- Data privacy: For scenarios in which users provide training data, attackers can repeatedly query a trained model to obtain users' private information.

3. Typical AI security attacks

3.1 Evasion

In an evasion attack, an attacker modifies input data so that the AI model cannot correctly identify the input. Evasion attacks are the most extensively studied attacks in academia. The following three types are representative.

Adversarial examples: Studies show that deep learning systems can be easily affected by well- crafted input samples, which are called adversarial examples. They are usually obtained by adding small perturbations into the original legitimate samples. These changes are not noticeable to human eyes but greatly affect the output of deep learning models. Szegedy et al. [2] first proposed adversarial examples in 2013. After that, scholars proposed other methods for generating adversarial examples. For example, the CW attack proposed by Carlini et al. can achieve a 100 percent attack success rate using small perturbations and successfully bypass most defense mechanisms.

Attacks in the physical world: In addition to perturbing digital image files, Eykholt et al. [3] describe modifying traffic signs to mislead an AI traffic sign recognition algorithm into identifying a "No Entry" sign as "45 km Speed Limit." In contrast to adversarial examples in the digital world, the physical world examples need to be scaling, cropping, rotation, and noise resistant.

Transferability and black-box attacks: To generate adversarial examples, attackers need to obtain AI model parameters, but these parameters are difficult to obtain in some scenarios. Papernot et al. [4] found that an adversarial example generated against a model can also deceive another model as long as the training data of the two models are the same. Attackers can exploit this transferability to launch black-box attacks without knowing AI model parameters. To do that, an attacker queries a model multiple times, uses the query results to train a "substitute model," and finally uses the substitute model to generate adversarial examples, which can be used to deceive the original model.

3.2 Poisoning

AI systems are usually retrained using new data collected after deployment to adapt to changes in input distribution. For example, an Intrusion Detection System (IDS) continuously collects samples on a network and retrains models to detect new attacks. In a poisoning attack, the attacker may inject carefully crafted samples to contaminate the training data in a way that eventually impairs the normal functions of the AI system—for example, escaping AI security detection. Deep learning requires a large number of training samples, so it is difficult to guarantee the quality of samples. Jagielski et al. [5] found that mixing a small number of adversarial examples with training samples can significantly affect the accuracy of AI models. The authors put forward three kinds of poisoning attacks: optimal gradient attack, global optimum attack, and statistical optimization attack. Demonstrations of these poisoning attacks targeted health care, loan, and house pricing data sets to affect the inference of AI models on new samples affecting dosage, loan size/interest rate, and house sale price predictions, respectively. By adding 8 percent of malicious training data, attackers can cause a 75-percent change of the dosages suggested by models for half of the patients.

3.3 Backdoor

Like traditional programs, AI models can be embedded with backdoors. Only the person who makes backdoors knows how to trigger them; other people do not know of their existence nor can trigger them. Unlike traditional programs, a neural network model only consists of a set of parameters, without source code. Therefore, backdoors in AI models are harder to detect than in traditional programs. These backdoors are typically implanted by adding some specific neuron into the neural network model. A model with a backdoor responds in the same way as the original model on normal input, but on a specific input, the responses are controlled by the backdoor. Gu et al. [6] proposed a method of embedding backdoors in AI models. The backdoors can be triggered only when an input image contains a specific pattern and, from the model, it is difficult to find the pattern or even to detect whether such a backdoor exists. Most of these attacks occur during the generation or transmission of the models.

3.4 Model extraction

In a model or training data extraction attack, an attacker analyzes the input, output, and other external information of a system to speculate on the parameters or training data of the model. Similar to the concept of Software-as-a-Service (SaaS) proposed by cloud service providers, AI-as-a-Service (AIaaS) is proposed by AI service providers to provide the services of model training, inference, etc. These services are open, and users can use open APIs to perform image and voice recognition. Tramer et al. [7] developed an attack in which an attacker invoked AIaaS APIs multiple times to steal AI models. These attacks create two problems. The first is the theft of intellectual property. Sample collection and model training require a lot of resources, so trained models are important intellectual property. The second problem is the black-box evasion attack mentioned earlier. Attackers can craft adversarial examples using extracted models.

4. AI Security layered defense

As illustrated in Fig. 1, three layers of defense are needed for deploying AI systems in service scenarios:

- Attack mitigation: Design defense mechanisms for known attacks.

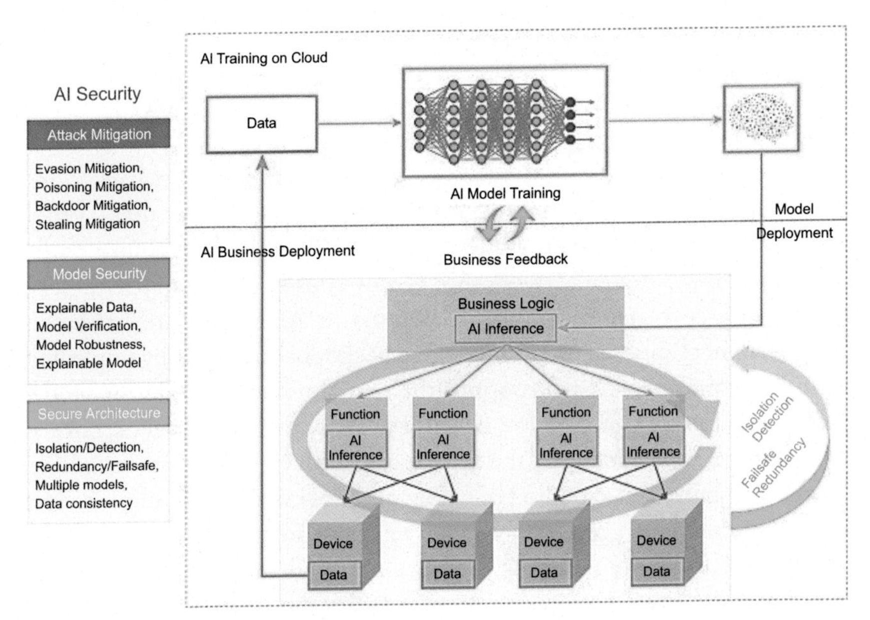

Fig. 1 AI security defense architecture.

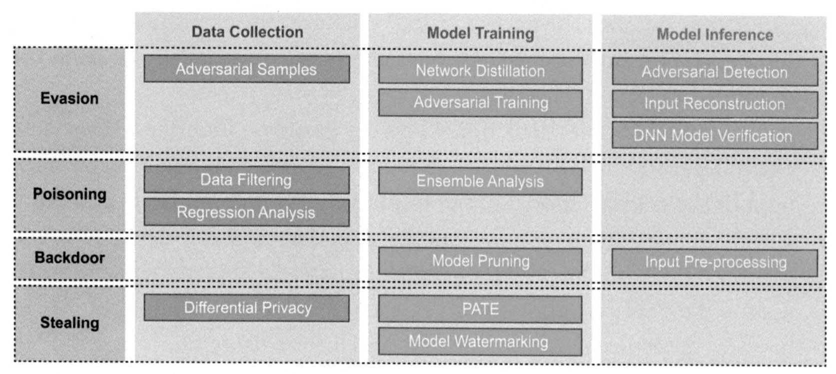

Fig. 2 AI security defense technologies.

- Model security: Enhance model robustness by various mechanisms such as model verification.
- Architecture security: Build a secure architecture with multiple security mechanisms to ensure business security.

Many countermeasures have been put forward in literature to mitigate potential attacks. Fig. 2 shows the various defense technologies used by AI systems during data collection, model training, and model inference.

4.1 Defense technologies for evasion attack

1. Network distillation: These technologies work by concatenating multiple DNNs in the model training stage, so that the classification result generated by one DNN is used for training the next DNN. Researchers found that the transfer of knowledge can reduce the sensitivity of an AI model to small perturbations and improve the robustness of the model [8]. Therefore, they proposed using network distillation technologies to defend against evasion attacks. Using tests on MNIST and CIFAR-10 data sets, they found that distillation technologies can reduce the success rate of specific attacks (such as Jacobian-based Saliency Map Attacks).
2. Adversarial training: This technology works by generating adversarial examples using known attack methods in the model training stage, adding the generated adversarial examples to the training set, and performing retraining one or multiple times to generate a new model resistant to attack perturbations. This technology enhances not only the

robustness but also the accuracy and standardization of the new model, since the combination of various adversarial examples increases the training set data.

3. Adversarial example detection: This technology identifies adversarial examples by adding an external detection model or a detection component of the original model in the model inference stage. Before an input sample arrives at the original model, the detection model determines whether the sample is adversarial. Alternatively, the detection model may extract related information at each layer of the original model to perform detection based on the extracted information. Detection models may identify adversarial examples based on different criteria. For example, the deterministic differences between input samples and normal data can be used as a criterion; and the distribution characteristics of adversarial examples and history of input samples can be used as the basis for identifying adversarial examples.
4. Input reconstruction: This technology works by deforming input samples in the model inference stage to defend against evasion attacks. The deformed input does not affect the normal classification function of models. Input reconstruction can be implemented by adding noise, de-noising, or using an automatic encoder (autoencoder) [9] to change an input sample. 5. DNN verification: Similar to software verification analysis technologies, the DNN verification technology uses a solver to verify attributes of a DNN model. For example, a solver can verify that no adversarial examples exist within a specific perturbation range. However, DNN verification is an NP-complete problem, and the verification efficiency of the solver is low. The efficiency of DNN verification can be improved by tradeoffs and optimizations such as prioritizing model nodes in verification, sharing verification information, and performing region-based verification.

These defense technologies apply only to specific scenarios and cannot completely defend against all adversarial examples. In addition to these technologies, model robustness enhancement can be performed to improve the resistance to input perturbations while keeping model functions to defend against evasion attacks. Multiple defense technologies can be combined in parallel or serially to defend against evasion attacks.

4.2 Defense technologies for poisoning attack

1. Training data filtering: This technology focuses on the control of training data sets and implements detection and purification to prevent

poisoning attacks from affecting models. This method can be implemented by identifying possible poisoned data points based on label characteristics and filtering those points out during retraining, or comparing models to minimize sample data exploitable in poisoning attacks and filtering that data out [10].

2. Regression analysis: Based on statistical methods, this technology detects noise and abnormal values in data sets. This method can be implemented in multiple ways. For example, different loss functions can be defined for a model to check abnormal values, or the distribution characteristics of data can be used.
3. Ensemble analysis: This technology emphasizes the use of multiple sub-models to improve an ML system's ability to defend against poisoning attacks. When the system comprises multiple independent models that use different training data sets, the probability of the system being affected by poisoning attacks is reduced.

An ML system's overall ability to defend against poisoning attacks can be further enhanced by controlling the collection of training data, filtering data, and periodically retraining and updating models.

4.3 Defense technologies for backdoor attack

1. Input pre-processing: This technology aims to filter out inputs that can trigger backdoors to minimize the risk of triggering backdoors and changing model inference results [11].
2. Model pruning: This technology prunes off neurons of the original model while keeping normal functions to reduce the possibility that backdoor neurons work. Neurons constituting a backdoor can be removed using finegrained pruning [12] to prevent backdoor attacks.

4.4 Defense technologies for model/data stealing

1. Private Aggregation of Teacher Ensembles (PATE): This technology works by segmenting training data into multiple sets in the model training stage, each for training an independent DNN model. The independent DNN models are then used to jointly train a student model by voting [13]. This technology ensures that the inference of the student model does not reveal the information of a particular training data set, thus ensuring the privacy of the training data.
2. Differentially private protection: This technology adds noise to data or models by means of differential privacy in the model training stage. For

example, some scholars propose a method [14] for generating gradients by means of differential privacy to protect the privacy of model data.

3. Model watermarking: This technology embeds special recognition neurons into the original model in the model training stage. These neurons enable a special input sample to check whether some other model was obtained by stealing the original model.

5. AI model security

As described above, adversarial ML exists extensively. Evasion attacks, poisoning attacks, and all kinds of methods that take advantage of vulnerabilities and backdoors are not only accurate, but also have strong transferability, leading to high risks of misjudgment by AI models. Thus, in addition to defense against known attacks, the security of an AI model itself must be enhanced to avoid the damage caused by other potential attacks. Fig. 3 illustrates some considerations on the enhancements.

5.1 Model detectability

Like traditional program analysis in software engineering, AI models could also be checked with some adversarial detection technologies such as black-box and white-box testing methods to guarantee some degree of security. However, existing test tools are generally based on open data sets having limited samples and cannot cover many cases in realistic deployments. Moreover, adversarial training technologies would cause high performance overhead due to re-training. Therefore, when AI models are being deployed in any system, a large number of security tests need to be performed on DNN models. For instance, a pre- processing unit could be used to filter out malicious samples prior to feeding into the training model, or a post-processing unit could be added to check the integrity of

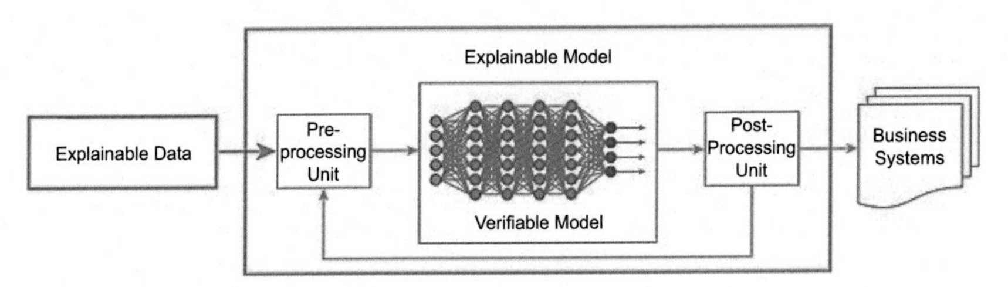

Fig. 3 Model security analysis.

the model output to further reduce false positives. With these methodologies, we could possibly enhance the robustness of AI systems before deployment.

5.2 Model verifiability

DNN models work surprisingly better than traditional ML techniques (for example, providing higher classification rates and lower false positive rates). Thus, DNN models are widely used in image and voice recognition applications. However, caution needs to be taken when applying AI models in security and safety sensitive applications such as autonomous driving and medical auto-diagnosis. Certified verification of DNN models can ensure security to some extent. Model verification generally requires restricting the mapping between the input space and output space to determine whether the output is within a certain range. However, since statistical optimization-based learning and verification methods typically cannot traverse all data distributions, corner cases like adversarial samples would still exist. In this situation, it is relatively difficult to implement specific protection measures in actual deployments. Principled defense can only be resolved when the fundamental working principle of DNN models is fully understood.

5.3 Model explicability

At present, most AI models are considered to be complicated black-box systems whose decision-making process, justification logic, and inference basis are hard to fully interpret. In some applications, such as playing chess and machine translation, we need to understand why machines make this or that decision to ensure better interaction between humans and machines. Nevertheless, the inexplicability of AI systems does not bring big problems in these applications. A translation machine can continue to be a complete complex black-box system as long as it provides good translation results, even though it does not explain why it translates one word into another. However, in some use cases, inexplicability tends to bring legal or business risks. For example, in insurance and loan analysis systems, if an AI system cannot provide the basis for its analysis results, it may be criticized as being discriminative; in healthcare systems, to accurately perform further processing based on AI analysis results, the basis of AI inference needs to be known. For instance, we hope that an AI system can analyze whether a patient has cancer, and the AI system needs to show how it draws the conclusion. Moreover, it is impossible to effectively design a secure model when its working principle is unknown. Enhancing the explicability of AI systems helps in analyzing their logic vulnerabilities or blind spots of data, thereby improving the security of the AI systems.

In the literature, researchers are actively exploring the explicability of AI models. For example, Strobelt et al. put forward the visualization analysis of hidden activation functions [15]. Morcos et al. proposed the use of statistical analysis methods to discover semantic neurons [16]. Selvaraju et al. offer saliency detection for image classification [17].Model explicability can also be implemented in three phases:

1. "Explainable data" before modeling: Because models are trained using data, efforts to explain the behavior of a model can start by analyzing the data used to train the model. If a few representative characteristics can be found from the training data, required characteristics can be selected to build the model during training. With these meaningful characteristics, the input and output of the model can be explained.
2. Building an "explainable model": One method is to supplement the AI structure by combining it with traditional ML. This combination can balance the effectiveness of the learning result and the explicability of the learning model and provide a framework for explainable learning. A common important theoretical foundation of traditional ML methods is statistics. This approach has been widely used and provides sound explicability in many computer fields, such as natural language processing, voice recognition, image recognition, information retrieval, and biological information recognition.
3. Explicability analysis of established models: This approach seeks to analyze the dependencies between the input, output, and intermediate information of AI models to analyze and verify the models' logic. In the literature, there are both general model analysis methods applicable to multiple models, such as Local Interpretable Model-Agnostic Explanations (LIME) [18], and specific model analysis methods that can analyze the construction of a specific model in depth.

When an AI system is explainable, we can effectively verify and check it. By analyzing the logical relationship between modules of the AI system and input data, for example, we can confirm that the customer reimbursement capability is irrelevant to the gender and race of the customer. The explicability of an AI system ensures clearer logical relationships between input data and intermediate data; this is the other advantage of AI system explicability. We can identify illegitimate or attack data, or even fix or delete adversarial examples based on the self- consistency of data to improve model robustness.

6. Security architectures of AI services

When developing AI systems, we must pay close attention to their potential security risks; strengthen prevention mechanisms and constraint conditions; minimize risks; and ensure AI's secure, reliable, and controllable development. When applying AI models, we must analyze and determine the risks in using AI models based on the characteristics and architecture of specific services, and design a robust AI security architecture and deployment solution using security mechanisms involving isolation, detection, failsafe, and redundancy.

In autonomous driving, if an AI system incorrectly decides on critical operations such as braking, turning, and acceleration, the system may seriously endanger human life or property. Therefore, we must guarantee the security of AI systems for critical operations. Various security tests are surely important, but simulation cannot guarantee that AI systems will not go wrong in real scenarios. In many applications, it may be difficult to find an AI system that can give 100 percent correct answers every time. This underlying uncertainty makes the security design of the system architecture more important. The system must enable fallback to manual operation or other secure status when unable to make a deterministic decision. For example, if an AI-assisted medical system cannot provide a definite answer about the required medicine and dosage or detects possible attacks, it is better for the system to reply "Consult the doctor" than to provide an inaccurate prediction that may endanger patient health. For safety's sake, correct use of the following security mechanisms based on business requirements is essential to ensure AI business security (Fig. 4):

1. Isolation: To ensure stable operation, an AI system analyzes and identifies the optimal solution and sends it to the control system for verification and implementation. Generally, the security architecture must isolate functional modules and setup access control mechanisms between modules. The isolation of AI models can reduce the attack surface for AI inference, while the isolation of the integrated decision module can reduce attacks on the decision module. The output of AI inference

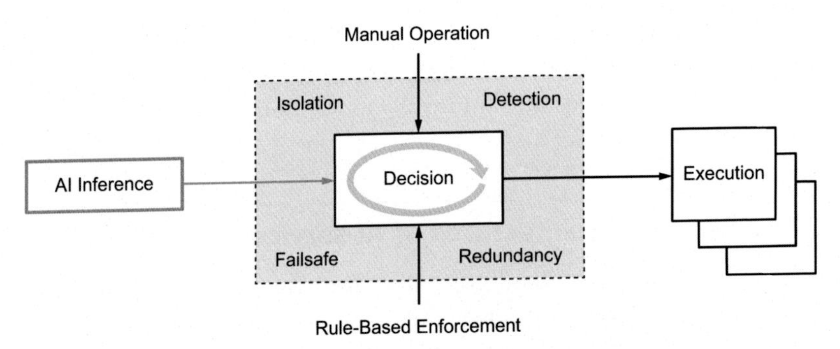

Fig. 4 Security architecture for AI with business decision making.

can be imported into the integrated decision module as an auxiliary decision-making suggestion, and only authorized suggestions can enter the decision module.

2. Detection: Adopting continuous monitoring and with an attack-detection model in the main system architecture, it is possible to comprehensively analyze the network security status and estimate the current risk level. When the risk is high, the integrated decision system can reject the suggestion coming from the automatic system and hand over control to a person to ensure security under attacks.
3. Failsafe: When a system needs to conduct critical operations such as AI-assisted autonomous driving or medical surgery, a multi-level security architecture is required to ensure the entire system security. The certainty of the inference results provided by the AI system must be analyzed. When the certainty of the result is lower than a certain threshold, the system falls back to conventional rule-based technologies or manual processing.
4. Redundancy: Many business decisions and data are associated with each other. A feasible method to ensure the security of AI models is to analyze whether the association has been ruined. A multi-model architecture can be set up for critical applications, so that a mistake in one model does not keep the system from reaching a valid decision. In addition, the multi-model architecture can largely reduce the possibility of the system being fully compromised by a single attack, thereby improving the robustness of the entire system.

Amodei et al. further describe a number of security challenges that AI systems may encounter during deployment [19]. For example, the authors describe how to avoid potential negative side effects during task execution and possible reward hacking during objective fulfillment, as well as safe exploration of AI systems. Fundamental research should be conducted on making AI systems more secure in future deployments.

7. Collaborating for a safe and a smart future

Various AI sub-disciplines, such as computer vision, voice recognition, natural language processing, cognition and reasoning, and game theory, are still in the early stages of development. Deep learning systems that leverage big data for statistical analysis have expanded the boundaries of AI, but they are also generally considered as "lacking common sense," which is the biggest obstacle to current AI research. AI could be both data driven and knowledge driven. The breakthrough of the next-generation AI lies in knowledge inference. Whatever the next steps, the large-scale popularization and advancement of AI require strong security assurance. This document focuses on two types of AI attacks and defenses.

First, affecting the correctness of AI decisions: An attacker can sabotage or control an AI system or intentionally change the input so that the system makes a decision desired by the attacker. Second, attackers may steal the confidential data used to train an AI system or extract the AI model. With these problems in mind, AI system security should be dealt with from three aspects: AI attack mitigation, AI model security, and AI architecture security. In addition, the transparency and explainability of AI are the foundation of security. An AI system that is not transparent or explainable cannot undertake critical tasks involving personal safety and public safety.

AI also requires security investigations in a wide range of domains, such as laws and regulations, ethics, and social supervision. On September 1, 2016, the One Hundred Year Study on Artificial Intelligence (AI100) project of Stanford University released a research report titled "AI and Life in 2030" [20], pointing out the profound changes brought by AI technologies and calling for regulatory activities that are more appropriate and that "will not stifle innovation." In the coming years, since AI will be deployed in the fields of transportation and medical care, the technology "must be introduced in ways that build trust and understanding, and respect human and civil rights." In addition, "policies and processes should address ethical, privacy, and security implications." To this end, international communities should cooperate to drive AI to evolve toward a direction that benefits mankind.

References

[1] I. Stoica, D. Song, R.A. Popa, D. Patterson, M.W. Mahoney, R. Katz, A.D. Joseph, M. Jordan, J.M. Hellerstein, J. Gonzalez, K. Goldberg, A. Ghodsi, D. Culler, P. Abbeel, A Berkeley View of Systems Challenges for AI, University of California, Berkeley, 2017. Technical Report No. UCB/EECS-2017-159.

[2] C. Szegedy, W. Zaremba, I. Sutskever, J. Bruna, D. Erhan, I. Goodfellow, R. Fergus, Intriguing properties of neural networks, arXiv (2013). preprint arXiv:1312.6199.

[3] K. Eykholt, I. Evtimov, E. Fernandes, B. Li, A. Rahmati, C. Xiao, A. Prakash, T. Kohno, D. Song, Robust physical world attacks on deep learning models, in: Conference on Computer Vision and Pattern Recognition (CVPR), 2018.

[4] N. Papernot, P. McDaniel, I. Goodfellow, Transferability in machine learning: from phenomena to black-box attacks using adversarial samples, arXiv (2016). preprint arXiv:1605.07277.

[5] M. Jagielski, A. Oprea, B. Biggio, C. Liu, C. Nita-Rotaru, B. Li, Manipulating machine learning: poisoning attacks and countermeasures for regression learning, in: IEEE Symposium on Security and Privacy (S&P), 2018.

[6] T. Gu, B. Dolan-Gavitt, S. Garg, Badnets: Identifying vulnerabilities in the machine learning model supply chain, in: NIPS MLSec Workshop, 2017.

[7] F. Tramèr, F. Zhang, A. Juels, M.K. Reiter, T. Ristenpart, Stealing machine learning models via prediction APIs, in: USENIX Security Symposium, 2016.

[8] N. Papernot, P. McDaniel, X. Wu, S. Jha, A. Swami, Distillation as a defense to adversarial perturbations against deep neural networks, in: IEEE Symposium on Security and Privacy (S&P), 2016.

[9] S. Gu, L. Rigazio, Towards deep neural network architectures robust to adversarial examples, in: International Conference on Learning Representations (ICLR), 2015.

[10] R. Laishram, V. Phoha, Curie: A method for protecting SVM classifier from poisoning attack, arXiv (2016). preprint arXiv:1606.01584.

[11] Y. Liu, X. Yang, S. Ankur, Neural trojans, in: International Conference on Computer Design (ICCD), 2017.

[12] K. Liu, D.-G. Brendan, G. Siddharth, Fine-pruning: Defending against backdooring attacks on deep neural networks, arXiv (2018). preprint arXiv:1805.12185.

[13] N. Papernot, A. Martín, E. Ulfar, G. Ian, T. Kunal, Semisupervised knowledge transfer for deep learning from private training data, arXiv (2016). preprint arXiv:1610.05755.

[14] M. Abadi, A. Chu, I. Goodfellow, H.B. McMahan, I. Mironov, K. Talwar, L. Zhang, Deep learning with differential privacy, in: ACM SIGSAC Conference on Computer and Communications Security, 2016.

[15] H. Strobelt, S. Gehrmann, H. Pfister, A.M.R. Lstmvis, A tool for visual analysis of hidden state dynamics in recurrent neural networks, IEEE Trans. Vis. Comput. Graph. 24 (1) (2018) 667–676.

[16] A.S. Morcos, D.G. Barrett, N.C. Rabinowitz, M. Botvinick, On the importance of single directions for generalization, arXive (2018). preprint arXiv:1803.06959.

[17] R.R. Selvaraju, M. Cogswell, A. Das, R. Vedantam, D. Parikh, D. Batra, Grad-CAM: visual explanations from deep networks via gradient-based localization, arXiv (2016). preprint arXiv:1610.02391.

[18] M.T. Ribeiro, S. Singh, C. Guestrin, Why should i trust you?: explaining the predictions of any classifier, in: ACM International Conference on Knowledge Discovery and Data Mining (KDD), 2016.

[19] D. Amodei, C. Olah, J. Steinhardt, P. Christiano, J. Schulman, D. Mané, Concrete problems in AI safety, arXiv (2016). preprint arXiv:1606.06565.

[20] S. Peter, B. Rodney, B. Erik, C. Ryan, E. Oren, H. Greg, H. Julia, K. Shivaram, K. Ece, K. Sarit, L.-B. Kevin, P. David, P. William, S. AnnaLee, S. Julie, Artificial Intelligence and Life in 2030, One Hundred Year Study on Artificial Intelligence: Report of the 2015–2016 Study Panel, Stanford University, 2016.

About the authors

Preetha Evangeline David is currently working as an Associate Professor and Head of the Department in the Department of Artificial Intelligence and Machine Learning at Chennai Institute of Technology, Chennai, India. She holds a PhD from Anna University, Chennai in the area of Cloud Computing. She has published many research papers and Patents focusing on Artificial Intelligence, Digital Twin Technology, High Performance Computing, Computational Intelligence and Data Structures. She is currently working on Multi-disciplinary areas in collaboration with other technologies to solve socially relevant challenges and provide solutions to human problems.

Anandhakumar is a professor in the Department of Information Technology at Anna University, Chennai. He has completed his doctorate in the year 2006 from Anna University. He has produced 17 PhD's in the field of Image Processing, Cloud Computing, Multimedia Technology and Machine Learning. His ongoing research lies in the field of Digital Twin Technology, Machine Learning and Artificial Intelligence. He has published more than 150 papers indexed in SCI, SCOPUS, WOS, etc.

CHAPTER SIX

Bayesian network-based quality assessment of blockchain smart contracts

K. Sathiyamurthy[a] and Lakshminarayana Kodavali[b]

[a]Department of CSE, Puducherry Technological University, Puducherry, India

[b]Department of CSE, Koneru Lakshmaiah Education Foundation (KLEF), Vadeswaram, Guntur, India

Contents

Abstract

Smart contracts (SC) are a key component in the Blockchain. Smart contracts are normal computer programs that are written mostly in solidity object-oriented programming language. Blockchain allows the completion of transactions only if the transaction obeying the smart contract rules. SC is not modifiable once they are deployed into the Blockchain. Thus it is essential to verify the quality of the smart contract before deploying it into the Blockchain. In this chapter, Bayesian Network Model is designed and constructed to measure the SC quality. The benefit of using a Bayesian network is

Advances in Computers, Volume 132
ISSN 0065-2458
https://doi.org/10.1016/bs.adcom.2023.07.004

that in addition to getting the probability of each quality measure, it will allow us to recommend the causes for the outcome of each quality measure. The proposed model with Bayesian Network helps in assessing the quality of smart contracts and the success rate is 82%.

1. Introduction

Smart contracts (SC) are the programs of predefined rules which are deployed into the Blockchain, and these programs execute automatically to determine that every transaction has to satisfy the predefined conditions to complete the transaction. In a Blockchain, the transactions among two parties are recorded in an efficient, verifiable, and unchangeable manner [1–4]. Blockchain can present an innovative solution for long-standing problems of security and data storage in the centralized systems [5]. Smart Contracts work based on conditional reasoning statements. Nowadays, smart contracts have been used widely in business among a group of untrusted persons, where every transaction can be completed according to rules agreed upon by all business stakeholders without the involvement of third-party verification [6]. Smart contracts have enabled the second generation of Blockchain technology to apply Blockchain to all other areas apart from cryptocurrencies [7].

The integration of SC with Blockchian technology causes vulnerability issues which leads to security attacks on SC. Smart contracts are written in Ethereum Blockchain using "Solidity," an object-oriented programming language and it is inspired by javascript, python and C++ languages. The causes of vulnerable smart contracts are unaware of SC vulnerabilities to the programmer, SCs can't be modified once they are deployed on to the Blockchain, and SC testing tools are still in the developing stage [8]. Hence, verifying the quality of smart contracts is essential and in contrast, smart contracts have no defined quality measures [9]. Thus, the proposed work provides a solution to the limitations of the SC quality measures using Bayesian Networks.

1.1 Importance of quality smart contract

Smart contracts can be used in many areas such as finance, land registrations, banking, insurance, energy, e-Government, telecommunications, voting systems, the music industry, healthcare, education and many more. The success of these applications with Blockchain depends on the quality of the smart contracts. The above-mentioned applications are very sensitive to data, prestigious, vital and money oriented. Blockchain is the transparent and permanent storage area for transactions but the success or failure of each transaction initiated by the customer depends on the smart contract. The SC allows the transaction to complete only if it satisfies the rules specified in the SC. There are more chances to attack, if the corresponding SC is vulnerable. It is very challenging in the current research to determine the correctness and security of smart contracts [10].

The literature proves that millions of dollars were lost due to SC security vulnerabilities [11], in different situations: (i) In 2016, due to DAO re-entrancy attack, there was a loss of $60 millions; (ii) In 2017, due to code injection attack, a loss of $30 millions; (iii) In 2018, due to integer overflow attack, indirect loss of billions of dollars. Many SC testing tools were developed but still there is more gap between detection effectiveness and programmer expectations [7,12]. All the existing tools can be accessible from GitHub but they do not provide good interface for effective use. The transaction related to writing into the Blockchain required to pay the gas but reading from Blockchain doesn't require to pay gas. These situations are forcing and motivating the researchers to develop the quality measures for smart contracts. Writing smart contracts intelligently without vulnerabilities in SC and to enable transaction speed is a challenging task. It requires expert knowledge about type of vulnerabilities and how to avoid them.

1.2 Software metrics

The Success of software development depends on writing the program in a simple manner and of high quality. There are different kinds of metrics for software quality that look at the different parts of the program such as its features, ease to use, meeting the customer's needs, getting rid of mistakes, making processes better and bringing in new users. The metrics are significant for the development and maintenance of software. The metrics help to make sure that the program works as per customer specifications and is designed in the right way. The software engineering team makes use of metrics to verify that the software meets all user requirements before releasing into the market. Metrics are also helpful to gather customer feedback about interface issues, bugs and errors, that have to apply for upcoming versions. In the literature [4,6] authors initiated the work towards measuring the quality of Blockchain smart contracts by developing tools (smart corpus [13], PASO [14] and SolMet [4]) to generate smart contract-related quality metrics as shown in Table 1.

Table 1 SC metrics extraction by different tools.

Smart corpus quality metrics	PASO quality metrics	SolMet quality metrics
SC address	Contract_definition	SLOC
Solidity version	Solidity version	LLOC
SLOC	SLOC	CLOC
CLOC	CLOC	NFNUMPAR
Mappings	Blanks	CC
Functions	Mappings	NL
Payable	Functions	NA
Events	Payable	**OO Metrics**
Modifier	Events	
Address	Modifier	DIT
	Address	CBO
	Contracts	NOA
	Libraries	NOD
	Interfaces	NOI
		NLE

1.3 Types of software quality measures

There are 10 important types of metrics that can be used to improve the quality of the software they are Dependable and Consistent, Functionality, Code Characteristics, Interface Accessibility, System Performance, Defect Management, Documentation, Adaptability across platforms, Safety & security and Overall satisfaction [15]. The proposed research work concentrates on six quality measures for determining quality of smart contracts which are code clarity (QM1), reliability (QM2), performance (QM3), code re-usability (QM4), maintainability (QM5) and security (QM6). Code quality metrics are the complexity, number of lines, number of functions and the bug generation rate. Reliability quality metrics are "Mean Time Between Failures" (MTBF) and "Mean Time to Repair" (MTTR). The performance quality metrics are time and resources being utilized. Code re-usability metrics represent how easy to integrate with other required software. Maintainability represents the time required to adapt new features or new functionality to existing software and MTTC ("Mean Time to Change") and to continue monitoring SC events to detect unexpected events. Security metrics represent there are no unauthorized changes and no threat of cyberattacks, when the software product is being used by the end user.

2. Literature work

In the literature, papers [1–5,16] were published on the smart contract quality metric. Nemitari Ajienka et al. [1] investigated the SCs nature in terms of their OO attributes to understand the relation with the GasUsed attribute. The authors [1] extracted the OO metrics which are CLOC, LLOC, SLOC, NF, NUMPAR, NL, CC, NLE, and NA using a tool "SolMet." The GasUsed attribute is more sensitive to the size measurements (SLOC, LLOC) and less to the structural characteristics (CBO or LCOM). The authors [1] stated, the attribute "GasUsed" has a significant correlation with the size attributes (NOS, LLOC and NL). The NOS (is part of the SLOC) and inheritance nesting levels (NL) metrics count must be low to reduce gas costs.

Tonelli et al. [2] collected the Blockchain addresses, the Solidity source code, the ABI and the byte code of 12,000 contracts from the Etherscan.io website and extracted smart contract-specific software metrics such as blank lines, CLOC, NF, SLOC, number of events calls, CC, number of mappings to addresses, number of payable, number of modifiers and perform analysis of which and to what extent the SC metrics influence smart contract performance. Their results proved that SC metrics have more restricted than the corresponding metrics in the traditional software system.

Giuseppe Antonio et al. [3] have noticed the availability of limited SC data sets and its metrics, authors developed the web-based repository [17] for SC's and its metrics for the last five years, it is easy to use, large, well-organized repository and very much helpful to users, researchers and developers of Ethereum Blockchain. Pierro and Tonelli [3] also provided a web-based SC repository (https://aphd.github.io/paso/), called PASO, which can provide SC commonly used metrics for the given smart contract address.

Matalonga et al. [4] focuses on the 15 security metrics which are common in both SC security metrics and general software security metrics. Authors [4] noticed that SC is security vulnerable if SC with high complexity in code structure. Hence SC developers have to take care SC complexity must be low and be simple as possible before deploying into Blockchain. The authors also stated that only the metrics WMC, DIT, CBO and CC can be collected using the "Solmet" tool and no tools are available to extract the remaining metrics which have to be collected manually. Auto extraction of these remaining metrics is one of the research issues.

Damian Rusinek et al. [5] proposed a 14-part checklist called SC Security Verification Standard (v1.1) to standardize the SC security at every stage of SC-DLC (Development Life Cycle) from design to implementation stages. This checklist will very much helpful to security reviewers, developers, architects, and vendors to avoid the majority of known security problems. Andrea Pinna et al. [16] collected 10K smart contracts from the Etherscan website and prepared meta-data regarding their interactions with Blockchain to understand the relationship between SC metrics and their impact on quality.

Authors [16] found that the SC metrics showed fewer values than traditional software metric values but had high variance. The authors stated different analysis results which are (i) found strong evidence on the practice of code reuse, that is new SC are developing by making use of existing SCs. (ii) SC name is not always matching with its actual work. (iii) number of transactions of SC showing power-law distribution, the SC with low balance may have many transactions and vice versa. (iv) number of transactions of an SC is not correlated to the SLOC of an SC that is well written SC. (v) SC complexity is mildly co-related to the number of functions (that is five times more than avg no. of functions per SC) and SLOC (greater than 300 lines). (vi) an average number of lines per SC is about 180 lines. (vii) SC are heterogeneous type and many deployed SC are deployed by inexperienced developers as a trail and experiment purpose without following standard SC structure.

Marco Comuzzi et al. [5] proposed the data carried by transaction payload plays a significant role in transaction execution in terms of time and cost. Data Quality (DQ) controls that are dependent on oracles (fetching off-line data) have a high impact on the number of required resources. DQ is evaluated by considering different DQ dimensions which are data consistency, metrics, completeness and accuracy. Each dimension can be assessed with multiple assessment metrics (single variable-single value, single variable-multiple values, multiple variables-single values, multiple variables- multiple values). Hence the authors suggest that Blockchain should also consider data controls while measuring its quality.

2.1 Related work on Bayesian networks

Bayesian Networks are significant for prediction or classification problems if we have prior probabilities of required events [18,19]. Daniel Kottke et al. [20] proposed a Bayesian approach to deal with uncertainties to determine posterior probabilities with help of prior distribution.

Eunjeong Park et al. [21] proposed a paper for predicting post-stroke outcomes with available risk factor probabilities using Bayesian Networks. The authors stated that Bayesian networks are preferable for complex and uncertain problems due to its interpretable capability than normal classification. The authors used hill climbing search technique to construct the Bayesian network. Meng et al. [22] proposed Bayesian network to evaluate risk factors versus uncertain demands in supply chain management and concluded that uncertain demands lead to high risk.

Zhao et al. [23] proposed Fuzzy theory in Bayesian networks to mine the required knowledge and data information from web text and present it in a way that users can easily understand. It is a challenging task to extract knowledge from web text which are organized using Bayesian networks, due to semi structure nature of web text and its complex data set. Chen et al. [24] stated and proved that probability techniques called Bayesian networks are suitable for complex engineering systems. Bayesian networks can process the complex systems effectively by reducing its complex data to reduced data (with major feature set) using probabilities. The authors used Min–Max Hill climbing learning algorithm for their Bayesian networks to predict event outcomes.

2.2 Gap in the literature

Generally, software quality measures are classified into code quality, reliability, code re-usability, maintainability, performance and security. In the literature the authors used tools (Smart Corpus, SolMet, and PASO) to extract few metrics of SC, which are not enough to measure all SC quality measures.

2.3 Motivation to choose proposed methodology

Bayesian Networks are preferred for this research work because they describe the severity (probability) of each quality measure, more interpretable, more scalable and they make us to describe the causes for each quality measure outcome (Low, Medium, High). After analyzing the limitations in the literature and the benefits of Bayesian Networks, the proposed work focused to use Bayesian Networks to measure the smart contract quality using the SC metrics and their impact on SC quality. The proposed work concentrates to measure the software quality in terms of code clarity, reliability, performance, code re-usability, maintainability and security. To the best of our knowledge, the usage of the proposed BSQM design is the first attempt to measure the smart contract quality using Bayesian networks. The next section described the architecture of the BSQM for SC Quality measures.

3. Smart contract quality metrics

In the literature authors used smart corpus, PASO [14] and SolMet [4] tools to generate smart contract-related quality metrics as shown in Table 1. The tool PASO generates many metrics compared with other tools. The abbreviations and definitions of the metrics listed in Table 1 can be got from the papers [4,9].

However, these limited metrics are not sufficient to meet all software quality measures. Hence the proposed BSQM model used additional metrics to evaluate all software quality measures (code clarity, reliability, performance, code re-usability, maintainability and security) as shown in Table 2.

In this research the BSQM model considered new metrics (external, require, revert, memory, import, SafeMath, transfer/call.value, uint8/16/../128, FP and SCLV) to measure the SC quality. The mapping between software quality measures and SC quality metrics is prepared after a deep analysis of software quality measures [15] and smart contract security verification standards [25–28] as shown in Table 3.

Code Quality Metrics: The SC quality measure "Code clarity" depends on the metrics CLOC, SLOC, and NF. The source code should have more number of simple functions (NF), more comments (CLOC) for each function to understand the purpose of the function and detailed source lines of code (SLOC).

Code Performance Metrics: The performance of an SC depends on the less time to complete each transaction, which in turn depends on the low gas usage that is monitored by the metrics: few number of state variables (NA),

Table 2 SC additional metrics proposed by BSQM model.

Software quality measures	Additional metrics proposed by BSQM
Code quality (**QM1**)	–
Reliability (**QM2**)	Require Revert
Performance (**QM3**)	Gas cost – memory – unit8/16/../128
Code reusability (**QM4**)	FP External Import
Maintainability (**QM5**)	Unit8/16/32/64/128 Memory Require Revert
Security (**QM6**)	SafeMath URC20 SCLV Transfer/call.value

Table 3 Mapping between software QM and SC metrics.

Software quality measures	Proposed SC quality metrics	Metric number
Code quality (**QM1**)	SLOC	M1
	CLOC	M2
	NF (functions)	M3
Reliability (**QM2**)	Mapping	M4
	Require	M5
	Modifier	M6
	Revert	M7
Performance (**QM3**)	Gas cost	
	– NA	M8
	– Memory	M9
	– unit8/16/../128	M10
Code reusability (**QM4**)	FP	M11
	External	M12
	Import	M13
Maintainability (**QM5**)	Event	M14
	unit8/16/32/64/128	M10
	Memory	M9
	Require	M5
	Revert	M7
Security (**QM6**)	SafeMath	M15
	SCLV	M16
	Payable	M17
	Address	M18
	Transfer/call.value	M19

many "memory" variables and many "uint8/16/32/64/128" variables. In SC programming any variable requires a minimum of 256 bits [29]. Hence to save memory and thereby gas consumption, it is preferable to combine multiple fewer storage variables into a single 256-bit length. It is possible to declaring a variables of type uint8, uint16, uint32, uint64 and uint128 to save memory space.

SC variables are of two types that are state variables and stateless variables. State variables can be write into the Blockchain to interact with it and they require gas. Number of state variables are also called as NA (number of attributes). NA value required to be less to save gas. State variables should not be declared as external. Stateless variables are local variables used for internal computations and do not require gas. Stateless variables can be declared with the "memory" keyword [29]. The metrics "int8/16/../128," "NA" and "memory" enable us to measure "code performance."

Code Reliability Metrics: The reliability of an SC depends on whether the SC is functioning as per restrictions specified in the SC which can be provided by the metrics that are mapping, require, modifier and revert. The function require() is allowed to proceed with execution flow further only if the specified condition in require() is met, otherwise it throws an error and stops execution. The function revert() is used to perform the revert operation explicitly in the methods. The metrics require(), revert() enable us to measure "code reliability".

Code Security Metrics: the security of SC can depend on the metrics which are the usage of "SafeMath" (safeAdd/safeSub/safeMul/safeDiv) to avoid arithmetic vulnerabilities, usage of solidity compiler latest version (SCLV) to avoid the limitation of older versions, usage of "transfer" or "call.value" functions along with a "payable" declaration to send money from one address to another address securely [30–33]. Pragma is a directive that specifies the solidity version to be used for the current SC. The solidity version can be specified as "pragma solidity 0.5.16" or "pragma solidity ^0.4.24." In the first example, SC can compile only with a specified compiler and in the second example, SC can be compiled with versions greater than 0.4.24 (called as FP—floating pragma). The metrics "SafeMath," "SCLV," "payable," "transfer," "call.value()" are enable us to measure "code security."

Code Reusability Metrics: The "code re-usability" of an SC depends on the metrics "import," floating pragma (FP) and the SC must declare functions with "external" otherwise, can't be accessed by other SC. The deployed SC with floating pragma can be easily compiled and compatible with other new SC. The already deployed SC can be used in newly developing SC with help of the "import" statement. The "external" keyword along with function declaration from deployed SC allowed to use further in new smart contracts. Hence the metrics "FP," "import" and "external" are considered to measure SC "code re-usability".

Code Maintainability Metrics: The maintainability of an SC depends on the (i) low gas usage for each transaction; (ii) continuous monitoring of the SC transaction logs to detect unexpected events which can be detected by the metrics "require" and "revert." The metrics "memory" and "uint8/16/../128" make SC to consume less gas, they enable to complete transactions quickly, thereby it is possible to continue the usage of SC without transaction failures. Hence above specified metrics are enable us to measure SC "code maintainability."

4. Proposed work

The architecture of the proposed work consists of two parts which are the BSQM design phase and the BSQM validation phase as shown in Fig. 1. The first phase (BSQM design) consists of mainly five modules which are data set collection, metrics extraction, preparation of metrics frequency table, creation of probability table and construction of Bayesian Network for each quality measure of each QM. The second phase (BSQM validation) of the architecture consists of a total of three modules which are metrics extraction for a given new SC, finding the probability of each QM using a probability table and providing BSQM results. The BSQM design phase is continued in this section. A detailed discussion about the BSQM validation phase is provided in Section 5.

4.1 Data set and metrics frequency table preparation

In the BSQM design phase, a data set of quality smart contracts are collected from the resources [34,35] by searching with the name SC domain name (Education/Healthcare) as shown in Table 4.

The data set in the current research consists of smart contract source code, how many transactions were performed for each SC, the address of SC, when it was deployed in Blockchain and other metadata of SC. Many SC transaction counts on this data set are showing mostly <5 that indicates these SCs are deployed by not experienced programmers and not been using further. Hence in the proposed research work, a data set is created with SCs which

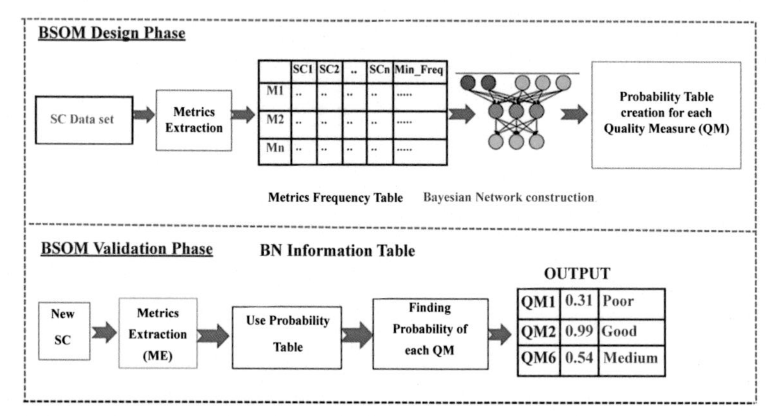

Fig. 1 The architecture of BSQM model for measuring smart contract quality.

Table 4 List of quality smart contract collected from the Etherscan.

S.no	Contract name	Address	Txion count	Published time
11	TutorcoinToken (Education)	0xe6c5036c814d180b3dea3578c75c c20510cf687d	2073	26/10/2018
22	CryptoRomeLandComposableNFT	0x86e4dc25259ee2191cd8ae40e186 5b9f0319646c	1591	04/09/2018
33	MortgageManager (LOAN)	0x9abf1295086afa0e49c60e95c437 aa400c5333b8	133	29/10/2018
44	MortgagePool	0xd8E5EfE8DDbe78C8B08bdd70A6dc 668F54a6C01c	71	07/05/2021
55	Employment (POEAToken)	0xa103f0fd3cfdf7dd6c4f98f61579 6f007112af3c	105	15/06/2020
6	Monthly Subscription (VRF Coordinator)	0x271682deb8c4e0901d1a1550ad2e 64d568e69909	72896	07/01/2022
7	Election (ElectionBetting)	0xf0037015bd137284f65b3842dd53 8ae204e32f2c	448	01/11/2020
8	OrbsVoting	0x30f855afb78758aa4c2dc706fb0f a3a98c865d2d	3837	25/03/2019
9	Voting (POOL)	0xcafea7934490ef8b9d2572eaefeb 9d48162ea5d8	3450	26/01/2021
10	FlashPoolV2	0x4c6997d462b0146fa54b74706541 1c1ba0248595	1845	20/08/2021

11	LandSale	0x7a11462a2adaed5571b91e34a127 e4cbf51b152c	1395	22/01/2019
12	Insurance (ThreeFMutual)	0x66be1bc6C6aF47900BBD4F371180 1bE6C2c6CB32	3129	19/09/2020
13	SmartContractBank	0x4f3db6d4C97b8248Ad5e244E4f61 9086084f6848	1498	26/10/2018
14	Government	0xf45717552f12ef7cb65e95476f21 7ea008167ae3	502	24/03/2016
15	Clinical (MixinMatchOrders.sol)	0x61935cbdd02287b511119ddb11ae b42f1593b7ef	892	17/11/2019
16	Supply Chain (LitionRegistry)	0x3b9a052bc3e457a0f278436f058e 040a147ab323	379	09/10/2019
...	......	......	....	....
100	OmnichainRouter	0x1de48d49aeba6ffa4740e78c84a1 3de8a9c12911	24	05/09/2022

Table 5 Smart contract metrics frequency table.

		Smart contract numbers					
S. no	**SC metrics**	**1**	**2**	**3**	**....**	**100**	**Min**
1	SLOC	118	542	50		1243	50
2	CLOC	10	25	15	...	38	10
3	NF	0	8	5	...	12	5
4	Mapping	5	15	14		7	5
5	Require	12	15	10		18	10
6	Modifier	5	22	14	...	8	5
7	Revert	4	6	3	...	0	3
8	NA	5	10	8		6	5
9	Memory	12	16	10	...	15	10
10	int8/16/32/../128	12	16	20	...	5	5
11	pragma ˆversion (FP)	Y		Y	...	Y	10 (Y)
12	External	3	67	39		5	3
13	Import	0	3	4		7	3
14	Event	7	0	5		6	5
15	SafeMath	4	6	8		5	4
16	SCLV	Y	N	Y	...	Y	8 (Y)
17	Payable	4	3	5	...	4	3
18	Address	4	5	0		3	3
19	Transfer/call.value	4	3	0	...	5	3

are having transaction count in between 500 and 4000. The SC metrics and their frequencies are collected for the data set using the resource PASO [14] as shown in Table 5. The last column in Table 5 shows the minimum value of the corresponding metric from the SC data set. In few SCs, the frequency of the metric could be 0, but it is not considered as a minimum, but MIN is considered from the values of greater than zero.

4.2 Probability table creation

This frequency table considers only the minimum frequency count to prepare the probability table, since max and average frequency values are varying based on the size of SC. The Probability Tables (PT) are created for each quality measure by taking into consideration of MIN frequency count (MFC) of each SC metric relevant to a particular QM as shown in Tables 6–11. MFC values can be get from Table 5 for each metric. After getting all MFC values for each metric in a particular quality measure, probability values can be calculated for each metric using Eqs. (1) and (2).

$$TFC_i = \sum_{m=1}^{n} MFC(m)_i \tag{1}$$

TFC_i = Total Frequency Count of $Q\mathbf{M}_i$
MFC_i = Min Freq Count of Metric $(m)_i$

$$\mathrm{Prob}(m)_i = \frac{MFC_m}{TFC_i} \tag{2}$$

Table 9 is showing PT for quality measure "code re-usability" (QM4). As per the Table 9 it is observed three metrics are influencing QM4 that are floating pragma (M11), external (M12) and import (M13) and hence Quality Measure List QML4 = {M11, M12, M13} or simply QML4 = {11, 12, 13}. Frequency count and Probability values for each QM can be get from Tables 6–12. According to Table 9, TFC is 10 + 3 + 3 = 16. The probability of M12 is 3/16 = 0.19. The metrics "floating pragma" (M11) and "SCLV" (M16) in Tables 9 and 11 respectively are marked with the "Y" symbol, indicating the answer for these two metrics is Yes (Y) or No (N). However, the MFC of these metrics is considered as per the number of SC in the data set having "Yes" for these metrics.

Table 6 Probability table for QM1.

QM1 metrics	MFC	Prob
M1	50	0.76
M2	10	0.15
M3	5	0.07

Table 7 Probability table for QM2.

QM2 metrics	MFC	Prob
M4	5	0.22
M5	10	0.43
M6	5	0.22
M7	3	0.13

Table 8 Probability table for QM3.

QM3 metrics	MFC	Prob
M8	5	0.25
M9	10	0.50
M10	5	0.25

Table 9 Probability table for QM4.

QM4 metrics	MFC	Prob
M11	10 (Y)	0.62
Ml2	3	0.19
Ml3	3	0.19

Table 10 Probability table for QM5.

QMS metrics	MFC	Prob
M5	10	0.30
M7	3	0.09
M9	10	0.30
M10	5	0.15
M14	5	0.15

Table 11 Probability table for QM6.

QM6 metrics	MFC	Prob
M15	4	0.19
M16	8 (Y)	0.38
M17	3	0.14
M18	3	0.14
M19	3	0.14

Table 12 Conditional probability table (CPT) for QM6.

M17	MI8	Prob(M19)
True	True	0.42
True	False	0.29
False	True	0.29
False	False	0.14

4.3 Bayesian network construction

Bayesian networks are a probabilistic graphical model that consists of nodes and directed edges between nodes. All variables/attributes are represented with nodes and conditional dependency between nodes are represented with directed edges. The missing connections between the nodes in the network indicates conditionally independent. BN models can be prepared by experts after careful analysis of data then the constructed model can be used to predict the test events. BN models can be challenging to design since lack of domain information completely to specify conditional dependence between variables. Even if available, it requires many calculations to specify full conditional probabilities for an event. Hence alternative solution is to specify dependencies between variables as per available data and treating remaining all the variables as conditionally independent.

In the proposed BSQM design, all metrics (Ex: M1, M2, ..., M19) are considered as nodes/circles and the sequence of edges between the nodes represents conditional dependencies between the metrics that are influencing a particular quality measure. All quality measures (Ex: QM1, ..., QM6) are represented as leaf nodes in the network. Bayesian Networks are prepared for each quality measure with help of Table 3 as shown in Figs. 2–7. These Bayesian networks are constructed after analyzing functional dependencies and sequences between metrics with respect to each quality measure as discussed in Section 3. Each node in the Bayesian networks is labeled with MFC and each edge is labeled with the probability of the corresponding metric in particular QM using probability tables. The probability value represents the influence of each metric on the corresponding quality measure.

4.4 Conditional probability table (CPT) preparation

Each node in the Bayesian network except leaf nodes consists of a Conditional Probability Table. The CPT for first-level nodes (from the top) which doesn't have parents, consists of only one entry (probability of itself).

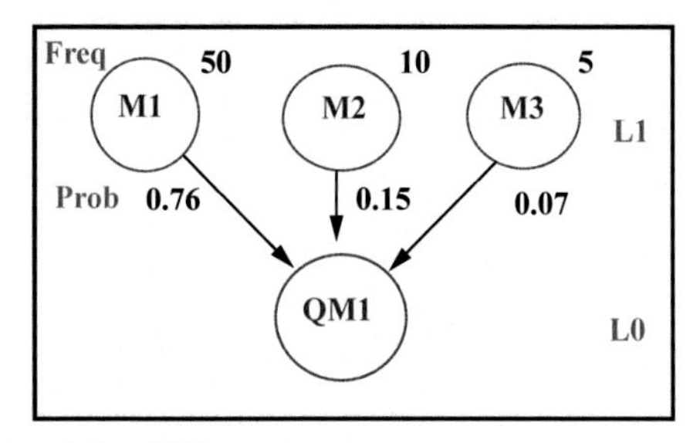

Fig. 2 Bayesian network for QM1.

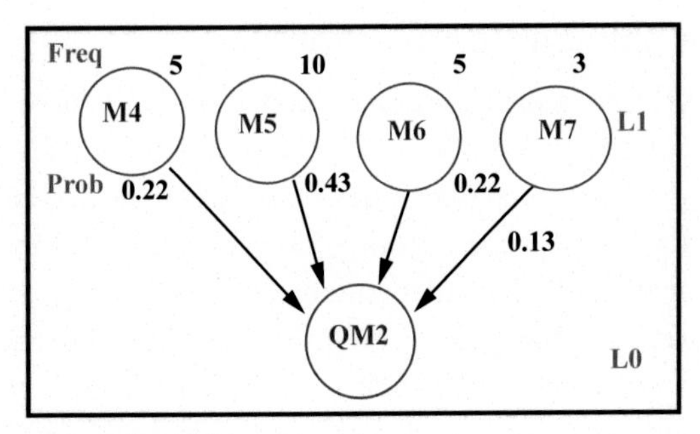

Fig. 3 Bayesian network for QM2.

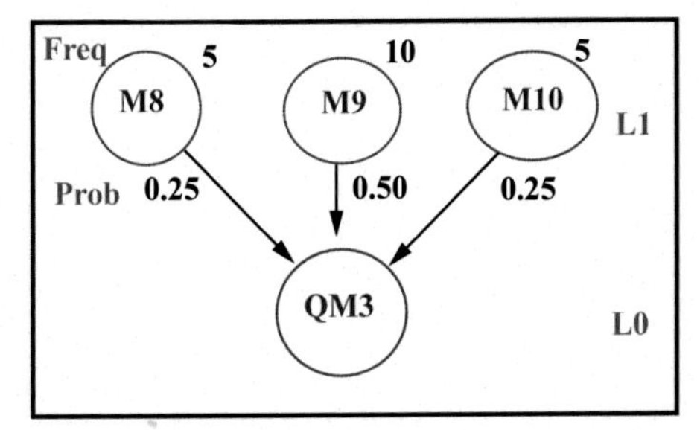

Fig. 4 Bayesian network for QM3.

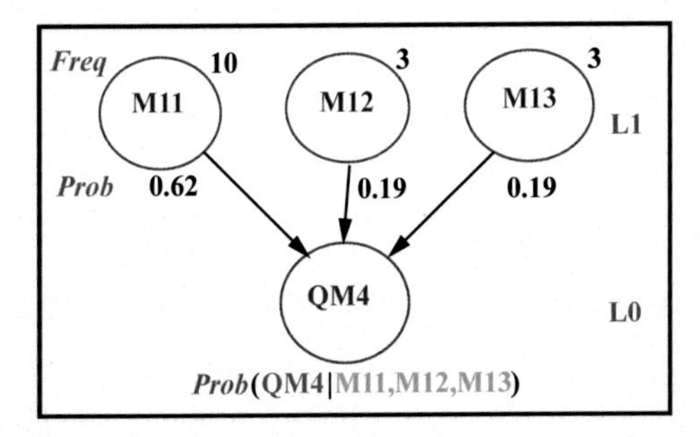

Fig. 5 Bayesian network for QM4.

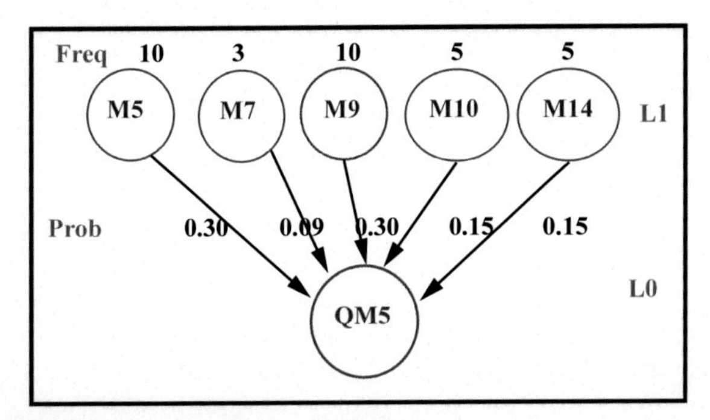

Fig. 6 Bayesian network for QM5.

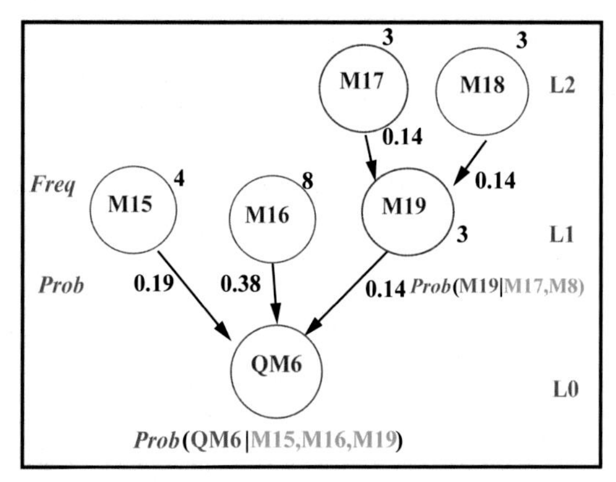

Fig. 7 Bayesian network for QM6.

For the dependent metrics (have parents) in the network, the number of entries in the CPT are 2^n, where n is the number of parent nodes on which the metric depends. The assignment of weight for the metric (m_i) in the network can be calculated using the Eq. (3). As per Eq. (3), all the nodes in the first level assigned with weight one (1) and weight will be keep on reducing as level increases in the BN.

$$W_{il} = \left\{ \frac{TL - CL}{TL - 1} \ if \ 0 < CL < TL \right\} \tag{3}$$

where TL = total levels, CL = current level of node

In Fig. 7, M19 (transfer) is conditionally dependent on the metrics M17 (payable) and M18 (address). It means the metrics M17 and M18 metrics along with M19 will influence more for the "code security" measure. The CPT for M19 is prepared using the Eq. (4) as shown in Table 12. The first row in the Table 12, represents the probability of the presence of M19 w.r.t presence of M17 and M18 is (9/21) = 0.42. The second row in Table 12 is the probability of the presence of M19 w.r.t the presence of M17 and the absence of M18 can be calculated as (6/21) = 0.29. The next section describes the experiment details and comparison results.

$$Cond.prob(m_i) = \frac{\sum (TrueMetricsFC)_i}{(TFC)_i} \tag{4}$$

TFC = total frequency count from the QM_i.
TrueMetricsFC = TM frequency count

5. Experimental and comparison results

The data set consists of smart contracts that are classified into vulnerability wise in separate documents. These smart contracts further classified as testing and testing purposes. In the BSQM validation phase, test smart contracts are collected from the resource [25] for testing purposes. The frequency Count (FC) for all metrics are extracted from the test SCs and applied algorithm as shown in Fig. 8 using Bayesian networks and probability tables for measuring the quality of smart contracts. The severity of each quality measure with respect to extracted metrics FC is examined for each SC.

In the BSQM validation algorithm, the extracted metrics FC from the testing SC are stored in the MFL variable in the form of [Metric Number: FC]. To calculate the probability of each quality measure (QM_i), it performs a sum of metrics probability (which are satisfying MFC) after multiplying with its weight, as shown in steps 6 and 7 in the algorithm as per Eq. (5). The DM list is used to store the metric numbers which are satisfying the MFC otherwise stored in the NDM list. The severity of each quality measure is displayed in a step 11 as per Eq. (6) and DM and NDM lists are printed in a step 12 as a proof of the severity of each QM.

SC_Quality_Measure_Algorithm(New_SC)

Input: New smart contract for testing its quality

Output: Displays severity of each quality measure and its causes

```
1.  MFL={Extract (Metrics: Frequency) from New SC}
2.  For i in [QM1, QM2, QM3, QM4, QM5, QM6]:
3.    DM=[], NDM=[] //Defined and Not defined Metrics
4.    For m in MFL {1:FC, 2:FC,..., n:FC} //Frequency Count
5.        If ( m in QML_i)
6.            If(FC_m>=MFC_i)
7.                Prob(QMi) += Prob(m) * Wm
8.                DM.append(m)
9.            Else
10.               NDM.append(m)
11.       print Quality severity of QM_i as per prob(QM_i)
12.       Print DM and NDM list for a proof of low/high severity
```

Fig. 8 Algorithm for BSQM validation. MFL, metrics frequency list; QML, quality measure list; DM, defined metrics; NDM, not defined metrics.

Table 13 Smart contract validation phase statistics for measuring its quality.

Metrics	1	2	3	4	5	6	7	8	9	10	11	12	13	14	15	16	17	18	19
FC	543	126	8	0	0	0	6	5	43	0	Y	0	7	0	0	N	4	6	3
Min. FC (MFC)	50	10	5	5	10	5	3	5	10	5	Y	3	3	5	4	Y	3	3	3
Satisfying count?	✓	✓	✓	×	×	×	✓	✓	✓	×	✓	×	✓	×	×	×	✓	✓	✓
Weight	1.0	1.0	1.0				1.0	1.0	1.0		1.0		1.0				0.5	0.5	1.0
Probability(m)	0.76	0.15	0.07				0.13	0.25	0.5		0.62		0.19						0.42
Prob(QM)	1.0			0.13				0.75			0.81			0.39	0.42				
Severity	High			Low				High			High			Med	Medium (Med)				
QM	Code Clarity			Reliability				Performance			Code Reusability			QM5	Security				

The statistics for a given tested SC's consists of extracted metrics FC, MIN count, probabilities, weights and its quality measure results are shown in Table 13. In Table 13, the Row-1 showing the metrics number, Row-2 showing extracted metrics FC for a given new SC, Row-3 showing the minimum frequency count (MFC) of each metric and Row-4 showing Yes/No for the metrics which are satisfying the MFC. The only metric which is satisfying the MFC will be considered for the probability calculation of each quality measure. The Row-5 showing the weight for all metrics that are satisfying the MFC as per Eq. (3).

$$prob(QMi) = \sum_{m \in QML} prob(mi) * Wm \tag{5}$$

$$where\ FCm > = MFCi(m)$$

$$QMseverity = \left\{ \begin{array}{lll} Low & if & 0.0 < prob(QM) < 0.29 \\ Medium & if & 0.3 < prob(QM) < 0.69 \\ High & if & 0.7 < prob(QM) < 1.0 \end{array} \right\} \tag{6}$$

Prob(QM1) = (0.76*1) + (0.15*1) + (0.07*1) = 0.98 (≈1.0).
Prob(QM5) = (0.13*1) + (0.5*1) = 0.39 (since M7, M9 are satisfying MFC from QM5 metrics list).
Prob(QM6) = (0.42*1) = 0.42 (it is taken from CPT of M19, with presence of M17, M18 metrics).

The output of the proposed BSQM model for the above-given tested SC's provides a probability for each QM, its severity (High, Medium, Low) along with causes of low or high probability in terms of defined and not defined metrics as shown in Table 14. In the BSQM validation phase, for the given tested SC's, the proposed model produced on an average 82% correct results compared with actual SC quality w.r.t High Severity in QM as shown in Fig. 9 and Table 15.

Table 14 Output from testing phase.

QM (quality measure)	Defined metrics	Not defined metrics	Probability (QM)	QM severity
Code clarity (QM1)	M1, M2, M3	—	1.0	High
Reliability (QM2)	M7	M4, M5, M6	0.13	Low
Performance (QM3)	M9	M8, M10	0.75	High
Code reusability (QM4)	M11, M13	M12	0.81	High
Maintainability (QM5)	M7, M9	M5, M10, M14	0.39	Low
QM grade (QM6)	M15, M16, M18,	M17, M19	0.64	Medium

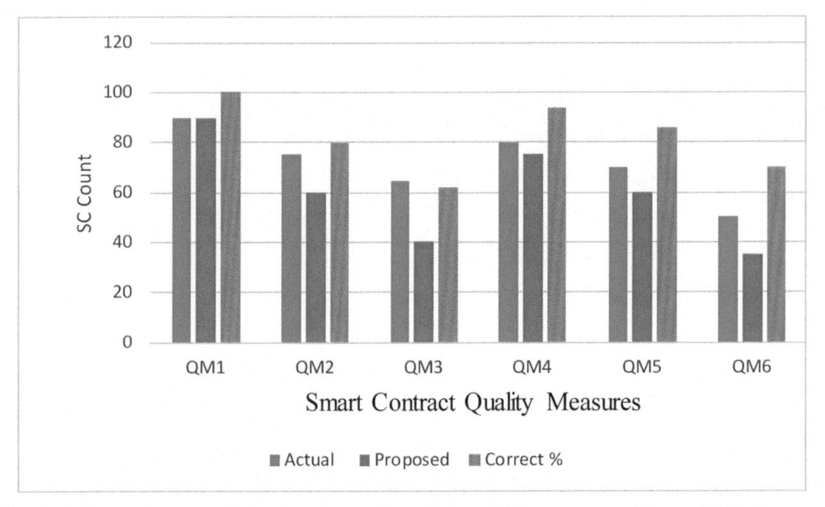

Fig. 9 Comparison graph between quality of actual SC vs proposed model QM.

Table 15 Comparison results between quality of actual SC vs proposed model QM.

	Actual	Proposed	Correct %
QM1	90	90	100
QM2	75	60	80
QM3	65	40	62
QM4	80	75	94
QM5	70	60	86
QM6	50	35	70

Table 16 Proposed BSQM model comparison with existing works.

	QM1	QM2	QM3	QM4	QM5	QM6
Nemitari Ajienka et al. [1]			✓		✓	
Tonelli et al. [2]			✓			
N'Da AAK et al. [4]	✓					✓
Damian Rusinek et al. [5]						✓
Andrea Pinna et al. [6]	✓			✓		
Proposed BSQM	✓	✓	✓	✓	✓	✓

The novelty of the proposed BSQM model can be observed from Table 16, which is focused on all six quality measures namely code quality, reliability, performance, code re-usability, maintainability and security. However, literature works have concentrated on only a maximum of two QM.

6. Conclusion

Smart contracts have great potential to apply for many real-life applications to adapt Blockchain. However, the main limitation of a smart contract is that there is a lack of well-defined quality metrics. Low-quality SC can't be modified once they are deployed into the Blockchain. Hence it is significant to check the quality of SC before deploying it into the Blockchain. In the proposed research work, the Bayesian network-based BSQM model is designed and implemented to measure the quality of smart contracts in terms of code clarity, reliability, performance, code re-usability, maintainability and security. The benefit of the proposed BSQM model is it enables us to provide inferences for high or low severity, in addition to providing the probability for each quality measure. The proposed BSQM design is implemented and evaluated on testing smart contracts and the results are demonstrating that, detecting correctly 100% for code clarity, 80% for reliability, 62% for performance, 94% for code re-usability, 86% for maintenance, 70% for security and overall success rate is 82%.

References

[1] N. Ajienka, P. Vangorp, A. Capiluppi, An empirical analysis of source code metrics and smart contract resource consumption, J. Softw. Evol. Process 32 (10) (2020) 1–22, https://doi.org/10.1002/smr.2267.

[2] R. Tonelli, G. Destefanis, M. Marchesi, M. Ortu, Smart contracts software metrics: a first study, arXiv (2018) 1802.01517 [cs.SE], no. February. https://doi.org/10.13140/RG.2.2.25506.12483.

[3] G.A. Pierro, R. Tonelli, M. Marchesi, An organized repository of ethereum smart contracts' source codes and metrics, Futur. Internet 12 (11) (2020) 1–15, https://doi.org/10.3390/fi12110197.
[4] N. Da, A.A. Kevin, UWS Academic Portal Applicability of the Software Security code Metrics for Ethereum Smart Contract Using Solidity UWS Academic Portal Applicability of the Software Security Code Metrics for Ethereum Smart Contract Using Solidity N'Da, Aboua Ange Kevin, 2021.
[5] M. Comuzzi, C. Cappiello, G. Meroni, An empirical evaluation of smart contract-based data quality assessment in ethereum, Lect. Notes Bus. Inf. Process. 428 (2021) 51–66, https://doi.org/10.1007/978-3-030-85867-4_5.
[6] K. Lakshmi Narayana, K. Sathiyamurthy, Automation and smart materials in detecting smart contracts vulnerabilities in Blockchain using deep learning, Mater. Today Proc. 81 (2021), https://doi.org/10.1016/j.matpr.2021.04.125.
[7] H. Rameder, M. di Angelo, G. Salzer, Review of automated vulnerability analysis of smart contracts on ethereum, Front. Blockchain 5 (2022) 1–20, https://doi.org/10.3389/fbloc.2022.814977.
[8] L.N. Kodavali, S. Kuppuswamy, Adaptation of blockchain using ethereum and IPFS for fog based E-healthcare activity recognition system, Trends Sci. 19 (14) (2022) 5072, https://doi.org/10.48048/tis.2022.5072.
[9] G. Antonio Pierro, R. Tonelli, in: PASO: a web-based parser for solidity language analysis, IWBOSE 2020—Proceedings of the 2020 IEEE 3rd International Workshop on Blockchain Oriented Software Engineering, 2020, pp. 16–21, https://doi.org/10.1109/IWBOSE50093.2020.9050263.
[10] L.B. Yiping, J. Xu, B. Cui, Smart Contract Vulnerability Detection Based on Symbolic Execution Technology, Springer Singapore, 2022, https://doi.org/10.1007/978-981-16-9229-1.
[11] K. Biswas, et al., Smart contract vulnerability detection model based on multi-task learning, Sensors 22 (2022) 1829, https://doi.org/10.3390/S22051829.
[12] SmartBugs, smartbugs/smartbugs: SmartBugs: A Framework to Analyze Solidity Smart Contracts. https://github.com/smartbugs/smartbugs. (accessed November 08, 2022).
[13] Smac-Corpus. https://aphd.github.io/smart-corpus/. (accessed October 29, 2022).
[14] PASO: SOlidity. https://aphd.github.io/paso/. (accessed November 09, 2022).
[15] Indeed.com, 10 Key Metrics for Software Quality and Why They Matter|Indeed.com, 2022. https://www.indeed.com/career-advice/career-development/metrics-for-software-quality. (accessed November 09, 2022).
[16] A. Pinna, S. Ibba, G. Baralla, R. Tonelli, M. Marchesi, A massive analysis of ethereum smart contracts empirical study and code metrics, IEEE Access 7 (2019) 78194–78213, https://doi.org/10.1109/ACCESS.2019.2921936.
[17] Smac-Corpus, 2022. https://aphd.github.io/smart-corpus/. (accessed November 04, 2022).
[18] D. Codetta-Raiteri, Bayesian networks: inference algorithms, applications, and software tools, Algorithms Spec. Issue 14 (5) (2021) 138, https://doi.org/10.3390/a14050138.
[19] T.M. Mitchell, Machine Learning, 1997.
[20] D. Kottke, M. Herde, C. Sandrock, D. Huseljic, G. Krempl, B. Sick, Toward optimal probabilistic active learning using a Bayesian approach, Mach. Learn. 110 (6) (Jun. 2021) 1199–1231, https://doi.org/10.1007/S10994-021-05986-9/TABLES/4.
[21] E. Park, H.J. Chang, H.S. Nam, A Bayesian network model for predicting post-stroke outcomes with available risk factors, Front. Neurol. 9 (2018) 699, https://doi.org/10.3389/FNEUR.2018.00699.
[22] M. Meng, Q. Lin, Y. Wang, The risk assessment of manufacturing supply chains based on Bayesian networks with uncertainty of demand, J. Intell. Fuzzy Syst. 42 (6) (2022) 5753–5771, https://doi.org/10.3233/JIFS-212207.

[23] Z. Wei, Z. Luo, Web text data mining method based on Bayesian network with fuzzy algorithms, J. Intell. Fuzzy Syst. 38 (4) (2020) 3727–3735, https://doi.org/10.3233/JIFS-179595.
[24] Y. Chen, T. Zhang, R. Wang, L. Cai, A Bayesian network structural learning algorithm for calculating the failure probabilities of complex engineering systems with limited data, J. Intell. Fuzzy Syst. 42 (3) (2022) 1991–2004, https://doi.org/10.3233/JIFS-211354.
[25] D. Rusinek, Smart Contract Security Verification Standards, [Online], Available: www.securing.biz.
[26] ConsenSys, Solidity Best Practices for Smart Contract Security|ConsenSys, 2022. https://consensys.net/blog/developers/solidity-best-practices-for-smart-contract-security/. (accessed November 09, 2022).
[27] 101 Blockchains, Solidity Smart Contract Security Best Practices—101 Blockchains. https://101blockchains.com/smart-contract-best-practices/. (accessed November 04, 2022).
[28] Getastra, Smart Contract Audit Services & Best Practices. https://www.getastra.com/blog/security-audit/smart-contract-security/. (accessed Nov. 09, 2022).
[29] Solidity Tutorial, 2022. https://www.tutorialspoint.com/solidity/index.htm. (accessed Nov. 09, 2022).
[30] Introduction to Smart Contracts, Solidity 0.8.17 documentation, 2022. https://docs.soliditylang.org/en/v0.8.17/introduction-to-smart-contracts.html. (accessed Nov. 09, 2022).
[31] S. Kalra, S. Goel, M. Dhawan, S. Sharma, ZEUS: analyzing safety of smart contracts, in: Network and Distributed Systems Security (NDSS) Symposium 2018, 2018, https://doi.org/10.14722/NDSS.2018.23082.
[32] J. Kim, M. Kim, Intelligent mediator-based enhanced smart contract for privacy protection, ACM Trans. Int. Technol. 21 (1) (2021), https://doi.org/10.1145/3404892.
[33] C. Bräm, et al., Rich specifications for Ethereum smart contract verification, Proc. ACM Program. Lang. 5 (2021) 30, https://doi.org/10.1145/3485523. OOPSLA.
[34] Etherscan.io, Ethereum Smart Contract Search, 2022. https://etherscan.io/searchcontractlist?a=all&q=education. (accessed November 04, 2022).
[35] klngithubsairam, klngithubsairam/BN_For_SC_Metrics: Bayesian Network for Smart Contract Metrics, 2022. https://github.com/klngithubsairam/BN_For_SC_metrics. (accessed November 09, 2022).

About the authors

Kuppuswamy Sathiyamurthy, was born in Tiruchirapalli, Tamil Nadu, India, on March 3, 1976. He has been working as a professor since 2002 at Puducherry Technological University, Puducherry, India. His special fields of interest included Machine Learning, Data Analytics, and Full Stack Technologies. He received a Doctor of Philosophy degree from Anna University, Chennai, India. He received his master's degree from Pondicherry University, Puducherry, India. He has published 37 international papers in reputed journals and conferences. He has also been involved in various software automation activities for administration and research projects.

Lakshminarayana Kodavali, was born in Renigunta, Andhra Pradesh, India, on May 29, 1982. His employment experience includes 15 years of teaching experience in the field of Computer Science and Engineering. His special fields of interest included Blockchain and Machine Learning. He completed his BTech and MTech from JNTU University Hyderabad, India. He is pursuing a PhD under the guidance of Dr. K. Sathiyamurthy from Puducherry Technological University, Puducherry, India. He has published 14 international papers in reputed journals and conferences.

CHAPTER SEVEN

Short-term wind power prediction using deep learning approaches

K.A. Alex Luke[a], Preetha Evangeline David[b], and P. Anandhakumar[a]

[a]Department of Information Technology, Madras Institute of Technology, Anna University, Chennai, Tamil Nadu, India

[b]Department of Artificial Intelligence and Machine Learning, Chennai Institute of Technology, Chennai, Tamil Nadu, India

Contents

Abstract

As the world's economy grows, the need for energy also increases. Traditional energy is being depleted, while environmental pollution is on the rise. This is why wind power is the most potential source of energy and a major renewable resource. Wind power generation is known as a widely-used and fascinating form of renewable energy generation around the world. However, the high uncertainty of wind power generation leads to some unavoidable errors in the wind power prediction process; consequently, it makes the optimal operation and control of power systems very challenging. Since wind power prediction error cannot be entirely removed, providing accurate models for wind power uncertainty can assist power system operators in mitigating its negative effects on decision making conditions. Conventional wind forecasting technologies are no longer adequate to effectively resolve the upcoming problems. As of now, the main solution to solve the problem is to forecast wind power in the future. By taking

Advances in Computers, Volume 132
ISSN 0065-2458
https://doi.org/10.1016/bs.adcom.2023.08.006

into account the different data available, deterministic predictions of wind power can be divided into statistical learning method predictions and physical model predictions. As a result of physical approaches, which are based on detailed descriptions of the atmosphere, weather data such as air temperature, topography, and pressure are used to predict wind speed. Here to predict the future wind energy utilization rate using machine learning algorithms such as time series analysis, neural networks, and support vector machines.

1. Introduction

Among the popular renewable energies, wind power technologies have been advanced and gained global attention as a result of the rapid expansion of the global energy demand. As a sustainable power resource for electric production, wind has been carefully examined and planned, especially with regard to minimizing the global warming effects. In recent years, wind power has grown to be one of the most significant renewable energy sources due to its benefits, which include reduced pollution, flexible investment, rapid construction, and minimal land occupation. Recently, with increasing usage of wind power plants worldwide, accurate and reliable power prediction is a challenging task. Forecasting of wind power production is useful for the energy industries. Wind power prediction makes full use of wind energy and it is important for the safe and stable power grid operation. Wind power prediction integrates large amounts of wind power into the electricity supply system and it also decreases the energy costs. With growing capacity there is also a growing demand for reliable and precise forecasts for the energy provided by the high volatile wind.

Wind power generation is the main source of the world's electricity generation, improving its prediction accuracy is important for making full use of wind energy and to ensure safe and stable operation of the power grid. Accurate forecasting of wind power generation is essential for the management and successful integration of large amounts of wind power into the electricity supply system. To decrease the energy costs, accurate prediction of wind power generation is important. Without forecasting, wind energy systems that are disorganized can cause irregularities and lead to great challenges to a power system [1]. So accurate prediction of wind power generation is essential. Optimization algorithms like whale optimization, Fruit fly Optimization, Ant Colony Optimization trapped into local optimum and failed to achieve Global optimum. So these Optimization algorithms can be enhanced by Chaotic Optimization Algorithm in order to achieve Global Optimum. Xin Li et al. [2] proposed a novel deep convolutional recurrent network method, the K-shape and K-means guided Convolutional Neural Network integrating Gated Recurrent Units (KK-CNN-GRU), for short-term multi-step ahead predictions of wind turbine power generation. The KK-CNN-GRU is composed of three modules, an input tensor construction module, a two-layer clustering module, and a prediction module.

Jordan Nielson et al. [3] proposed a robust machine learning methodology used to generate a site-specific power-curve of a full-scale isolated wind turbine operating in an atmospheric boundary layer to drastically improve the power predictions, and, thus, the forecasting of the monthly energy production estimates. The ANN model in this study uses Feed Forward Back Propagation (FFBP) algorithm. Guoqing et al. [4] proposed a wind power prediction model that combines Adaboost algorithm with extreme learning machine optimized by particle swarm optimization (PSO-ELM). First, particle swarm optimization is used to optimize the initial thresholds and input weights of the ELM to obtain the PSO-ELM basic prediction model. Then, combined with the Adaboost algorithm, a series of PSO-ELM weak predictors with input weights and thresholds optimized by PSO and containing different hidden layer nodes are composed. Yimei Wang et al. [5] proposed a probabilistic WPF model which is established based on the probabilistic characteristics of Gaussian Process, the summation of different kernels is used to model the variations of output wind power with wind speed and wind direction. The established GPR model can accurately describe the fluctuation process of wind power by taking wind speed and wind direction as input, and has better generalization ability than RBF NN. Cao et al. [6] proposed a chaotic ant colony optimized (CACO) link prediction algorithm, which integrates the chaotic perturbation model and ant colony optimization. The extensive experiments on a wide variety of unweighted and weighted networks show that the proposed algorithm CACO achieves significantly higher prediction accuracy and robustness.

Wind power prediction (WPP) is based on different time horizons which categorize into very-short-term, short-term, medium-term and long-term predictions [7]. Very Short term forecasting refers to predicting wind power before 30 min. Short term wind power prediction refers to predicting the wind power before 1 to 12 h. Medium term Forecasting refers to predicting the power before 12 to 72 h. Long term forecasting refers to predicting the power before 2 weeks. For a sustainable integration of wind power into the electricity grid, a precise and reliable Short term prediction method is required. There are two big challenges for applying the deep learning techniques. First, the prediction error and reliability have to be improved. Second, the required computation time needs to be reduced. By filtering the noise, accuracy has been improved and computation time reduced. The wind power measurements contain similar features of noise which have contributions in time and frequency [4] which highly affects the power prediction. Meta-heuristic algorithms are used to optimize only the key parameters of the prediction model stage, and the other stages are rarely studied [8]. This makes it impossible to identify the stage in which a large prediction error is caused by the combined forecasting framework [8]. The effective selection of the features and the high-accuracy prediction model are the main challenges in power prediction. This paper aims at predicting the wind power by filtering out the noise effectively.

2. Dataset and components of wind turbine

The dataset contains various weather, turbine and rotor features. Readings have been recorded at a 10-min interval. The dataset consists of 21 attributes. Those attributes includes ActivePower, Ambient Temperature, Bearing ShaftTemperature, Blade1PitchAngle, Blade2PitchAngle, Blade3PitchAngle, Control box Temperature, Gearbox Bearing Temperature, Gearbox Oil Temperature, GeneratorRPM, GeneratorWinding1Temperature, Generator Winding2Temperature, Main Box Temperature, Nacelle Position, Reactive Power, Rotor RPM, Turbine Status, Wind Turbine Generator (WTG), Wind Speed, Wind Direction, Hub Temperature. The active power refers to the power that continuously flows from source to load in an electric circuit. In an electric circuit, reactive power constantly flows from a source to a load before returning to the source. Active power is called Real power or True power or Actual power. Blade pitch angle refers to the angle between the propeller blade chord line and the plane of rotation of the propeller. From the Fig. 1, it is inferred that the components Low speed shaft, generator, nacelle position, hub, pitch are taken as a parameter based on these parameters wind power is predicted.

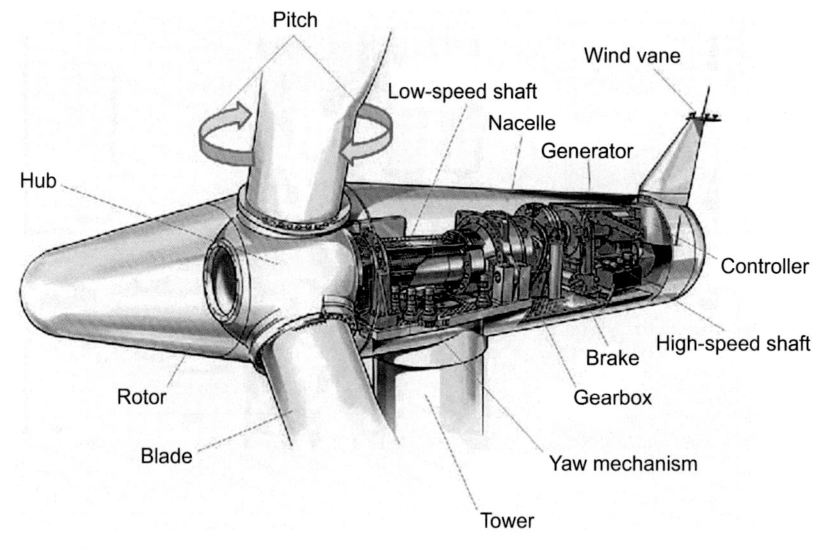

Fig. 1 Wind turbine components.

3. Proposed work

The architecture focuses on the data obtained from the dataset and the flow represents to preprocess the data and to identify noise level and to filter out the noise. From the Fig. 2 represents the flow of predicting the wind power.

The proposed wind power prediction mainly includes Data Pre-Processing, Feature Selection, Model Training and Power Prediction.

Step 1: Data Pre-Processing

Data Pre-Processing is performed using RobustScaler, Standard Scaler in order to identify the missing values.

Step 2: Feature Selection

The Features are selected using Random forest, Decision tree and Linear Regression algorithms.

Step 3: Model Training

The models are planned to be constructed using Support Vector Regressor/GRU model/Convolutional Neural Network/LSTM

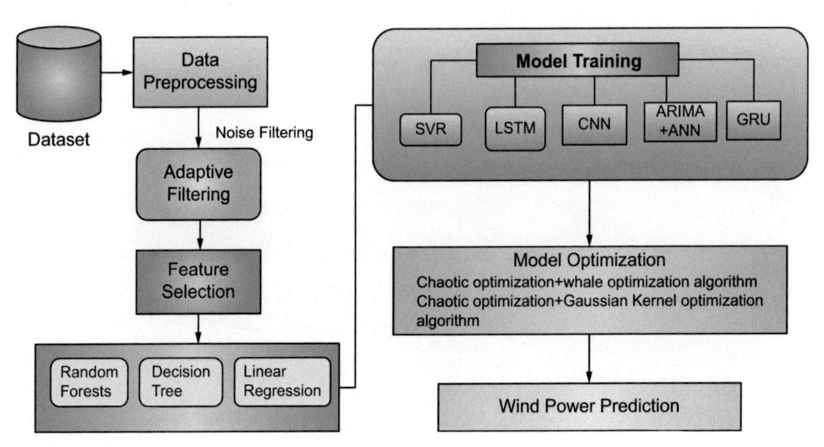

Fig. 2 Proposed architecture.

followed by Optimization algorithms like Gaussian Kernel Optimization/ Whale Optimization algorithms in order to achieve high accuracy in power prediction.

3.1 Data preprocessing

As a part of data preparation, data preprocessing refers to any type of processing done on raw data to get it ready for another data processing technique. It has long been regarded as a crucial first stage in the data mining process.

3.1.1 Robust scalar

Statistics are used to scale features that are robust to outliers. This technique scales the data between the first and third quartiles while removing the median. That is, between the 25th and 75th quantiles. This range is also called an interquartile range.

The median and interquartile range are then saved to be applied to new data using the transform method in the future. If outliers are present in the dataset, then the median and the interquartile range provide better results and outperform the sample mean and variance [9].

$$\frac{x_i - Q_1(x)}{Q_3(x) - Q_1(x)} \tag{1}$$

3.1.2 Standard scalar

Standard Scaler helps to obtain a standardized distribution with a zero mean and standard deviation of unit variance. By subtracting the mean value from the feature, and then dividing the result by the standard deviation, features are standardized. The Eqs. (1) and (2) are the formulas of robust scalar and standard scalar respectively which is used to preprocess the data. The data is analyzed and read from the dataset.

The formula for standard scaling is:

$$z = (x - u)/s \tag{2}$$

After analyzing the data preprocess it to find the missing values and outliers which are not related to the Wind Power Prediction. On finding the missing values the missing values are filled with the median of the feature columns. Then check for null values in the attributes and there is no null values in of the dataset. The correlation between the attributes are found in order to find the highly correlated attributes which plays a major role in power prediction.

From the Fig. 3, it is inferred that the dark color represents highly correlated attributes and light color represents low correlated attributes. The value ranges of the map is from 0 to 1 [10]. Those features which are highly correlated have their value as 1. From the map, can see that the nacelle position and wind direction are highly correlated, and the generator RPM and rotor RPM are highly correlated. So, that it can ignore one of the features from the two highly correlated features. The data is preprocessed using robust scaler, standard scaler.

3.1.3 Gaussian kernel optimization

Gaussian Process Regression is helpful to forecast targets from features on a time-series scale, however it hugely depends on the used similarity kernel whether to fit or not. So, when one has limited data or needs small-scale forecasts from the close past, Gaussian performs well [11].

- Constant kernel

  Can be used as part of a product-kernel where it scales the magnitude of the other factor (kernel) or as part of a sum-kernel, where it modifies the mean of the Gaussian process [12].

- ExpSineSquared kernel

  The ExpSineSquared kernel allows one to model functions which repeat themselves exactly. It is parameterized by a length scale parameter $l > 0$ and a periodicity parameter $P > 0$. Only the isotropic variant where l is a scalar is supported at the moment.

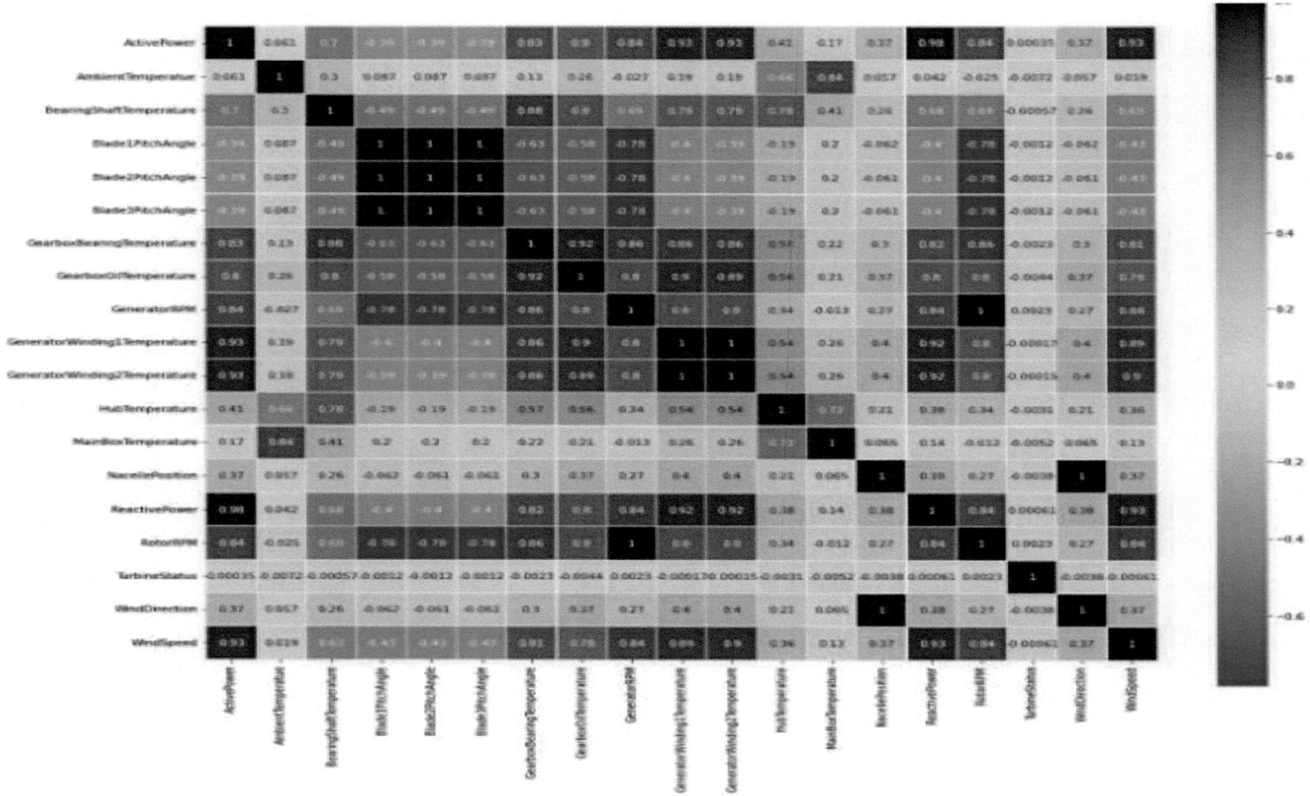

Fig. 3 Visualization using heatmap.

- White Kernel

  The main use-case of this kernel is as part of a sum-kernel where it explains the noise of the signal as independently and identically normally-distributed.
- Radial Basis Function

  The RBF kernel is a stationary kernel. It is also known as the "squared e exponential" kernel. It is parameterized by a length-scale parameter $l > 0$, which can either be a scalar (isotropic variant of the kernel) or a vector with the same number of dimensions as the inputs x (anisotropic variant of the kernel).

The distributions of various derived values can be computed directly when a random process can be represented as a Gaussian process [13]. The Pickle function is used for serializing and de-serializing the object structures. Based on the four kernels has to do with all possible combinations and has to adjust with the hyperparameters. First, the kernel functions needs to be defined then compose the kernels. To increase iterations of Gaussian Process Regressor to overcome the problem synchronization optimization which are not inherited in post processing techniques. The problem is solved by using Gaussian Process Regressor. The kernels which are not suitable for data are suppressed. To filter the features of noise which have contributions in Time and Frequency, highly affects the wind power measurements. Adaptive Kernel Filter is used to filter the noise. After filtering out the noise, the data is analyzed.

3.2 Feature selection

3.2.1 Feature selection using random forest

Random forest is a supervised machine learning technique that can be used for classification and regression. It builds decision trees from several samples and uses the majority vote for classification and the average for regression [14]. The basic workflow of the random forests is shown in Fig. 4. The random forest approach has the advantage of being able to handle data sets with both continuous and categorical variables, as in regression and classification. It eliminates decision tree overfitting. It has built in wrapper and filter methods. The attributes in the dataset is fed into different classifiers [14]. The classifiers then select the features and based on voting mechanism the features are selected.

Random Forest selects 19 features and their performance metrics are calculated. The get_support function returns an array of true or false values. Here, the array returns the true values which represents the selected features.

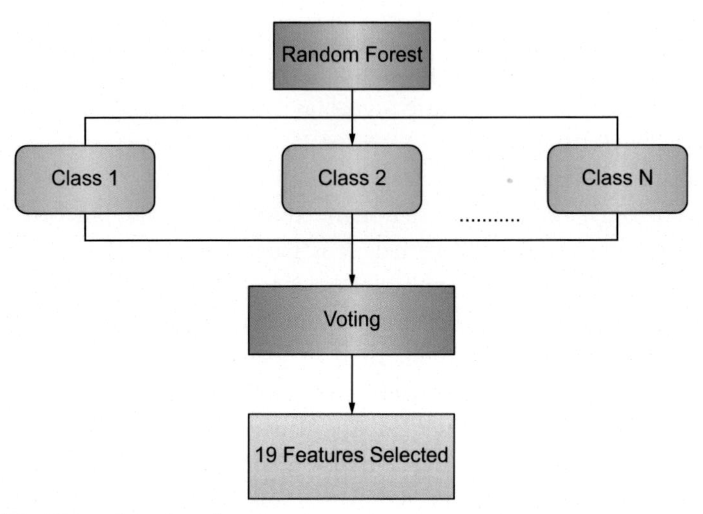

Fig. 4 Workflow of random forests.

The len function returns the length of the selected features. The features are selected using random forest models and the MSE, RMSE, MAE values are determined.

3.2.2 Feature selection using decision tree regressor

Decision trees can be used for selecting features or screening variables completely. They can handle both categorical and numerical data. Each internal node of the tree representation denotes an attribute and each leaf node denotes a class label. In addition, they can handle multiple outputs and results [15]. It breaks down a dataset into smaller and smaller subsets while at the same time an associated decision tree is incrementally developed.

By learning simple decision rules from the features of the data, are attempting to create a model that can predict the value of a target variable. Decision trees are constructed that identify different conditions under which a data set can be splitted into training and testing dataset and the features are selected. New features can also be created for better target variable prediction. Decision trees can handle both numerical and categorical data and are not greatly impacted by outliers or missing values. Since it is a non-parametric method, it has no assumptions about space distributions and classifier structure. The workflow of the decision tree regressor is shown in Fig. 5.

The decision tree regressor model is used to select the best features. Here, the column_list contains the set of all attributes of the dataset.

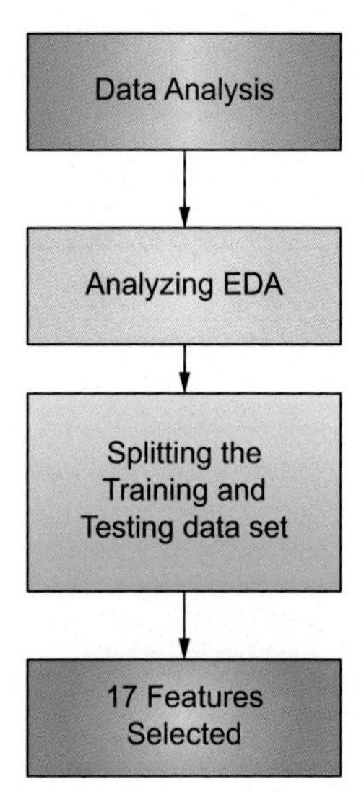

Fig. 5 Workflow of decision tree regressor.

The active power is given as the target parameter and selected the features. Based on the target parameter i.e., active power, the scores for other parameters are calculated. The MSE, RMSE, MAE, MAPE, SMAPE are determined for this model.

3.2.3 Feature selection using linear regression

A supervised machine learning algorithm is Linear Regression. It carries out the task of regression. Based on the independent variables, regression models a goal prediction value.

It's usually used to figure out how variables relate to one another and to forecast [16]. Its job is to predict the value of a dependent variable (y) given a set of independent variables (x). As a result, a linear relationship between input and output is discovered using this regression technique.

Linear regression is used to perform feature selection. The column_list contains the set of all attributes of the dataset. Here, the target parameter is set as the active power and selected the features with respect to the target parameter. With respect to the target parameter i.e. Active power, the scores for other parameters are calculated. The MSE, RMSE, MAE, MAPE, SMAPE are determined for this model.

The comparison is made between three feature selection models in Table 1. On comparing the metrics of these models, Decision Tree

Table 1 Performance analysis of feature selection.

Algorithms	MSE	RMSE	MAE	Features selected
Decision Tree Regressor	0.077	0.122	0.077	17
Linear Regression	0.15	0.23	0.15	15
Random Forests	3.89	191.18	3.89	19

Regressor outperforms [29] other models by selecting 17 features with minimum error rates. The selected features are then given to the model for further processing.

3.3 Model training and optimization

3.3.1 Support vector regressor

Support Vector Regression (SVR) is a regression algorithm that can handle both linear and nonlinear regressions. It's used to forecast ordered continuous variables. SVR's main goal is to find the optimum fit line [17,18].

The support vectors assist in determining the closest match between the data points and the function that represents them [19,20]. GridSearchCV technique is used to search through the best parameter values from the given set of the grid of parameters. SVR model and the parameters are required to be fed in then the best parameter values are extracted and then the predictions are done [21,22].

To obtain the best parameters for SVR and use the best parameters directly and do the residual analysis. By analyzing the result, it could see that this method works quite well. The RMSE is drastically reduced, as well as the mean error, so that the prediction is better overall. The correlation metric shows that the predictions are close to the actual values [23]. The fundamental idea behind SVR is to nonlinearly map the original data onto a high-dimensional feature space in order to achieve the best generalization capabilities.

The SVR model is used to predict the generation of wind power. The model is trained and tested with real-time time series data and the metrics are calculated. Here the X-axis is the Time and Y-axis the Power. Fig. 6 represents the actual and the predicted wind power generation predicted by the SVR model. The blue line indicates the actual or true wind power whereas the orange line indicates the predicted wind power of the model.

Artificial Bee Colony Algorithm

1. Initialization Phase
2. REPEAT

```
In [38]: plot_random_predictions(pred_power_svr, actual_power_svr)
```

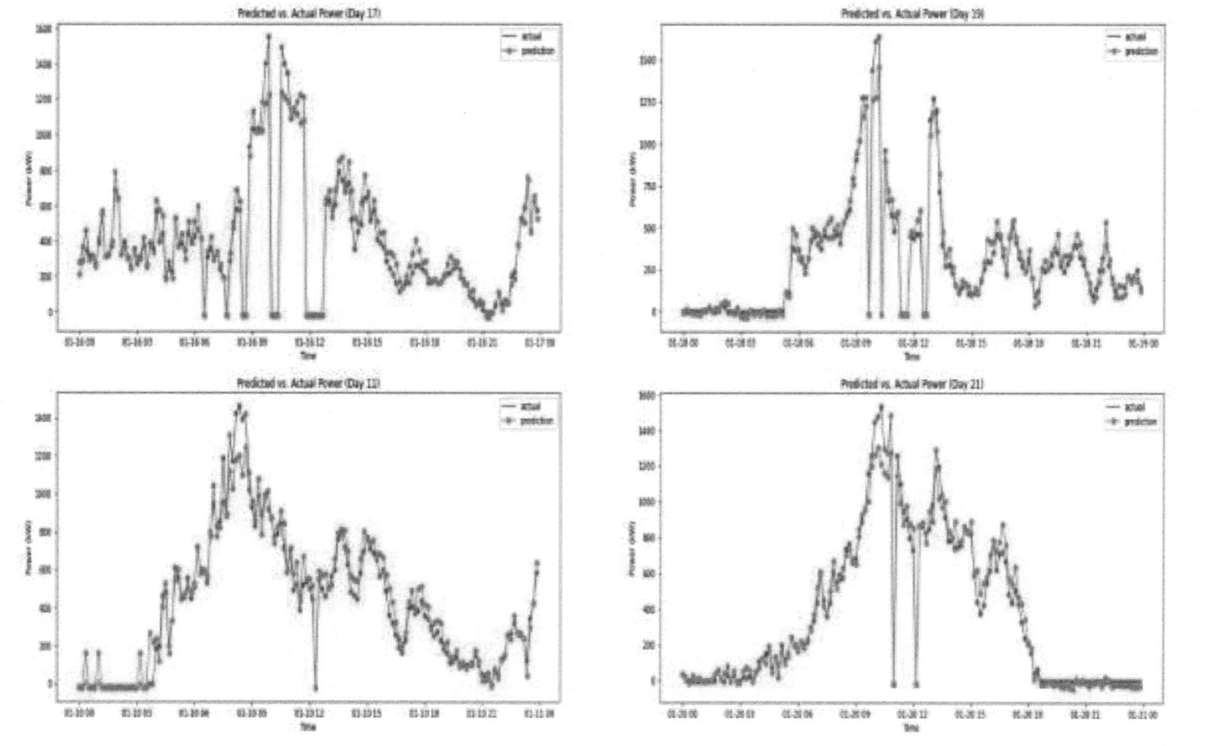

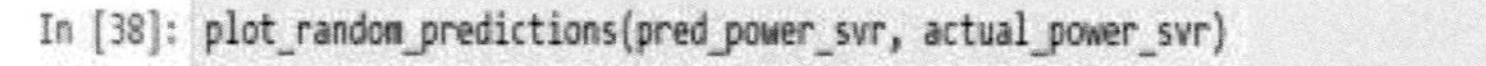

Fig. 6 Actual power Vs Predicted power using SVR model.

3. Employed Bees Phase
4. Onlooker Bees Phase
5. Scout Bees Phase
6. Memorize the best solution achieved
7. UNTIL(Cycle = Maximum Cycle Number or a Maximum CPU time)
ABC algorithm is used for optimization. Reconstructing the phase space of the System needs to examine every component. By constructing an m-dimensional vector from the component and its fixed delay point observations, the original observation sample space can be mapped to an equivalent high-dimensional state space. Optimizing the initial position of the source parameter by uniform distribution to speed up the convergence speed of the algorithm. Due to the uncertainty of the structure and parameters of the neural network, the prediction interval is uncertain, which is the main factor for the uncertainty of wind power prediction. To overcome this problem, the SVR [24] model is optimized with Artificial Bee Colony. After performing Iterative Optimization the optimal neural network parameters are fed into the model.

After performing optimization with ABC the results achieved are very close to the actual results and the error rate is also minimized which is shown in Fig. 7.

3.3.2 Long-short term memory

LSTM stands for Long short-term memory. Recurrent neural networks that learn to predict the future from sequences of varying lengths use LSTM cells. It is capable of handling the vanishing gradient problem faced by RNN. Unlike any feedforward neural network, LSTM has feedback connections [25]. As a result, it can predict values for both point and sequential data. LSTMs became popular because they could solve the problem of vanishing gradients [26]. As it turns out, they are unable to entirely remove it. The fact that the data still needs to be transferred from cell to cell for analysis is the difficulty. LSTMs get affected by different random weight initialization and hence behave quite similar to that of a feed-forward neural net [27]. LSTMs are prone to overfitting and it is difficult to apply the dropout algorithm to curb this issue. They require a lot of resources and time to get trained and needs high memory-bandwidth [28]. LSTM isn't able to learn as GRU, since it has more parameters and it is concluded that it is not that efficient.

The metrics for the lstm model is calculated. The actual and the predicted wind power determined by the lstm is shown in the Fig. 8. After optimizing the lstm model with ABC optimization algorithm, the results are shown in the Fig. 9.

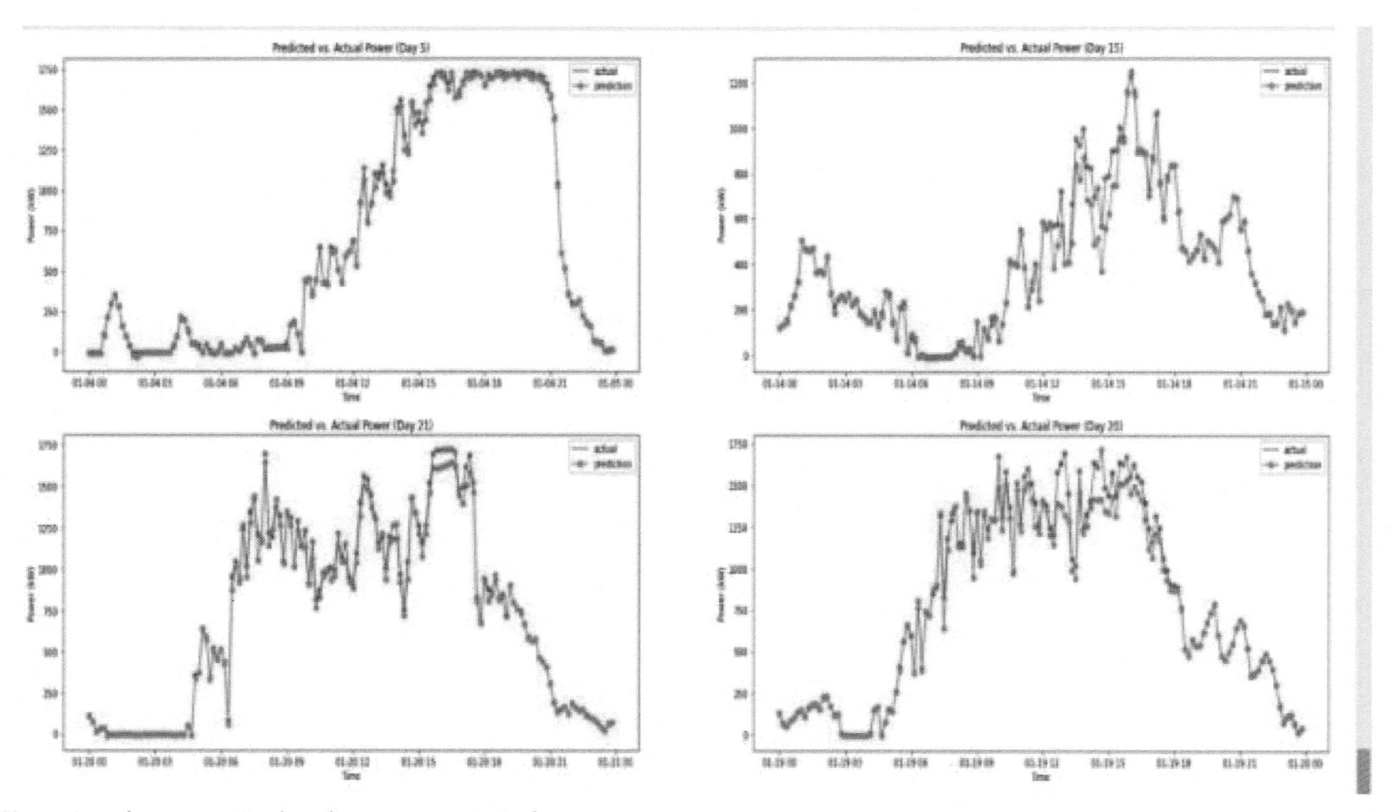

Fig. 7 Actual power Vs Predicted power using SVR after ABC optimization.

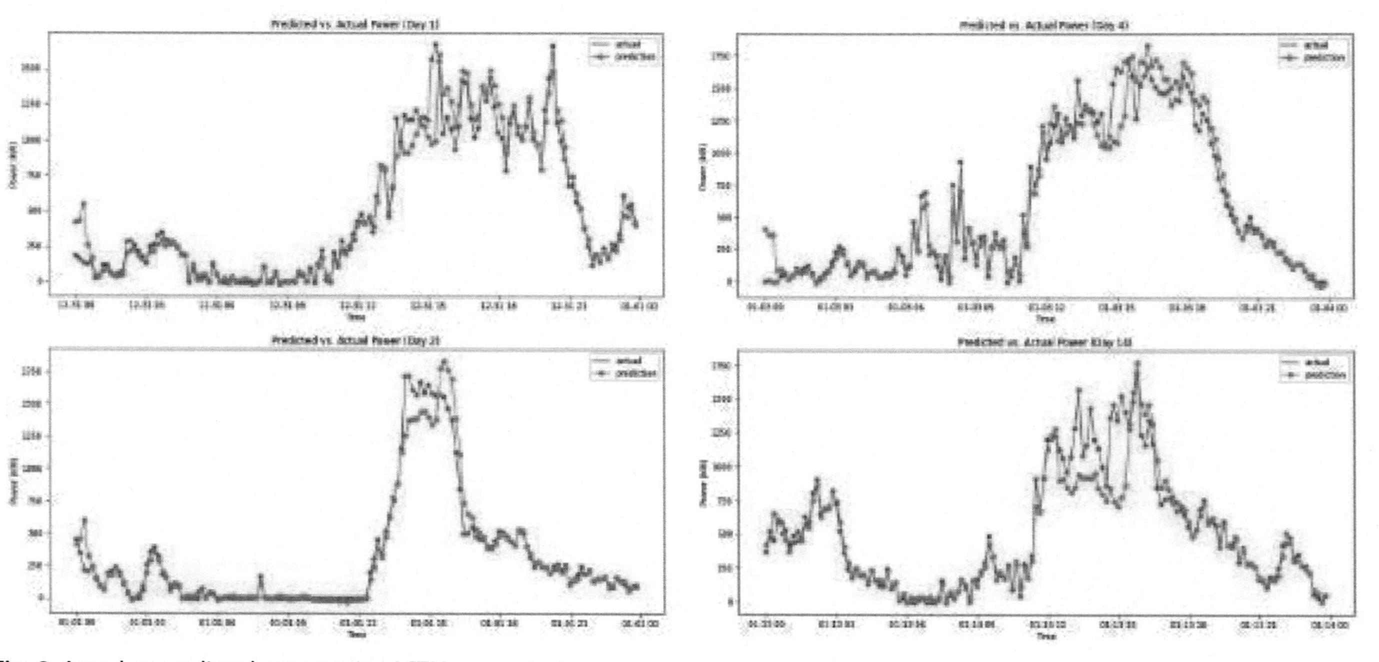

Fig. 8 Actual vs predicted power using LSTM.

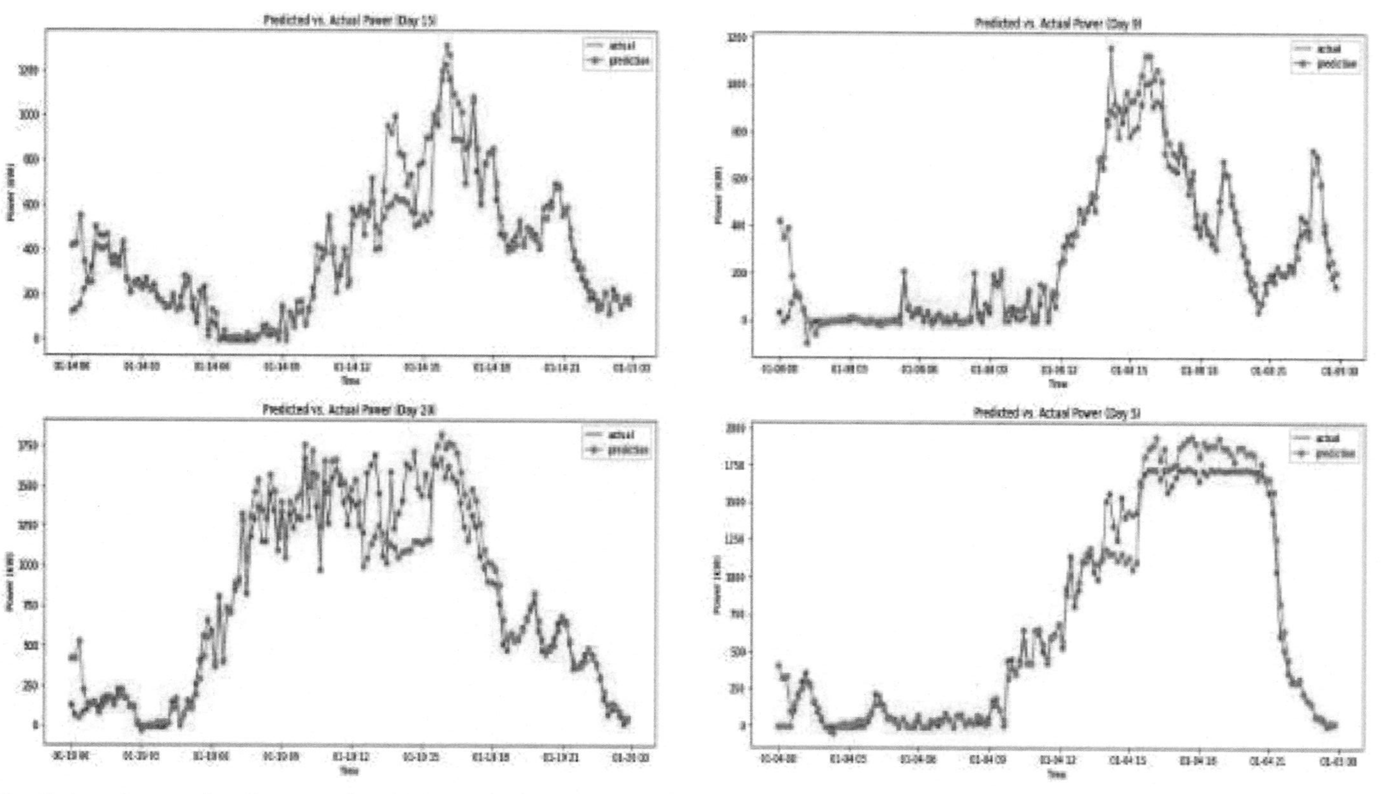

Fig. 9 Actual vs predicted power after ABC optimisation using LSTM.

3.3.3 Gated recurrent unit

The gated recurrent unit is a sort of LSTM-based recurrent neural network that is optimized. The GRU internal unit is similar to the LSTM internal unit, except that the GRU merges the incoming and forgetting ports into a single update port, whereas the LSTM does not. It consists of only three gates and does not maintain an Internal Cell State. The LSTM immunity to the vanishing gradient problem is preserved. Its underlying structure is simpler, making it easier to train because upgrading the internal states requires less math [29]. GRUs address the vanishing gradient problem from which vanilla recurrent neural networks suffer [30]. Gated RNNs can better capture dependencies for sequences with large time step distances. Splitting and reshaping are done. After validation, the model captures the data well, but cannot reduce the error as well as the SVR model. The metrics for the gru model is calculated. The actual and the predicted wind power determined by the gru model is shown in the Fig. 10. After optimizing the gru model with ABC optimization algorithm, the results are shown in the Fig. 11.

3.3.4 Convolutional neural network

The main applications of a convolutional neural network (CNN), which comprises one or more convolutional layers, are image processing, classification, segmentation, and other autocorrelated data. The main advantage of CNN compared to its predecessors is that it automatically detects the important features without any human supervision. Here Gramian Angular Fields to convert the time series to image of size *nxn* being *n* the number of features. As a result, these images undergo non-Cartesian transformations whenever a pixel is pointed at. Considering that CNNs are quite powerful, maybe they are able to improve the score over RNNs [31].

To reobtain the x_train, x_val and x_test, reshaping them to 4 dimensions to fit the input size and then running the parameters. Then comparing the actual values with the predicted values of CNN. Then Optimizing the parameters using ABC optimization technique and comparing it with actual and predicted values. It seems that after optimization the error rate is dropped. The actual and the predicted wind power determined by the CNN model is shown in the below Fig. 12.

The results after optimization with ABC are shown in the Fig. 13. Although after optimization with the ABC algorithm, CNN isn't learning well. It may be that the dimensionality of the input data is too low for a CNN, making the learning difficult for such a number of parameters.

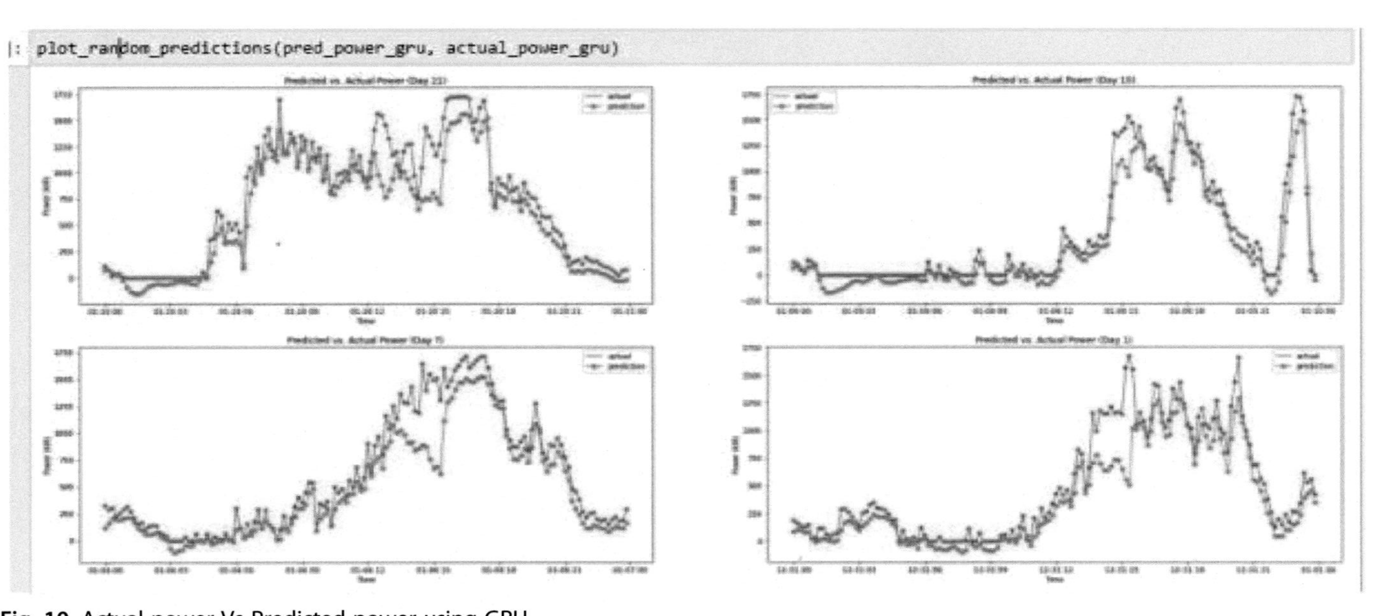

Fig. 10 Actual power Vs Predicted power using GRU.

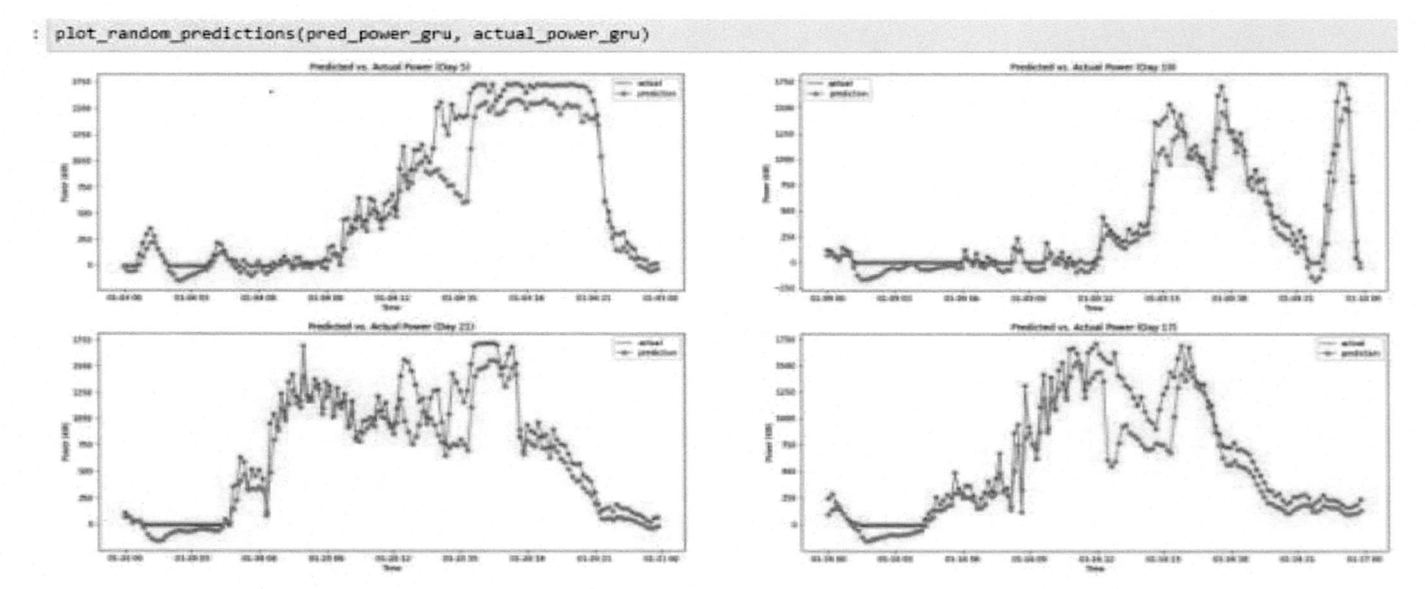

Fig. 11 Actual power Vs Predicted power after ABC optimisation using GRU.

```
In [26]: plot_random_predictions(pred_power_cnn, actual_power_cnn)
```

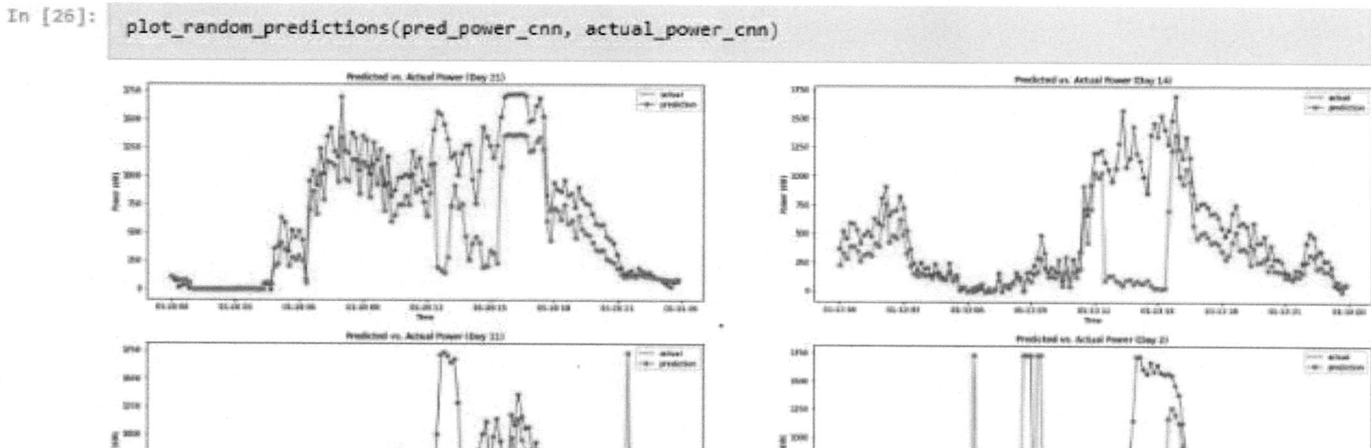

Fig. 12 Actual power Vs Predicted power using CNN.

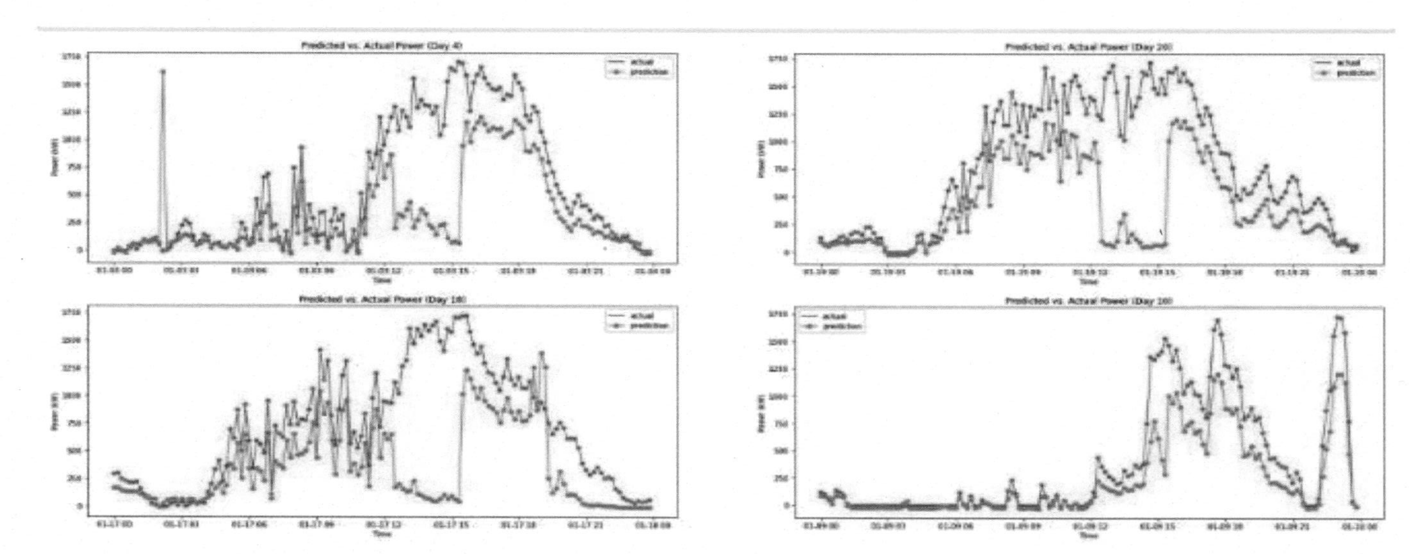

Fig. 13 Actual power Vs Predicted power after ABC optimisation using CNN.

Table 2 Performance analysis of models.

Algorithms	MAE	RMSE
SVR	39.70	68.4
GRU	88.71	212.73
LSTM	142.565	237.03
CNN	275.094	460.827

Three models are used to predict the generation of wind power. When these models are compared with each other, SVR shows minimum error rates and predicts the generation of wind power comparatively better than other models which has been shown in the Table 2.

Further, optimization algorithms are applied for SVR models to make the model perform even better.

3.3.5 Chaotic particle swarm optimization for SVR model

The chaotic search method is to use its own ergodic characteristics to search and find the best solution in this way [32]. One of the major drawbacks of the PSO is its premature convergence, especially while handling problems with more local optima. Particles tend to be stagnant when their velocities are near to zero. A real-valued encoding scheme based on a matrix representation is developed, which converts the continuous position value of particles in PSO to the processing order of job operation. A compound chaotic search strategy that uses the Logistic chaotic search process is employed to the global best particle to enhance the local searching ability of PSO [33]. The metrics after optimizing the SVR model with CPSO are calculated. In the PSO algorithm, a swarm of particles continuously updates a search space, and at the conclusion, an ideal solution is found.

Randomly initialize a population of particles with positions and velocities. In the optimized issues, the locations stand in for potential solutions.

- Evaluate the fitness value for each particle.
- Determine the location of each particle's ideal fitness value based on its previous movement. The best position of the nth particle is denoted.
- Determine where the best fitness value is for each particle. The best position for all particles.
- Particle position and speed should be updated.

Logistic function—an "S" shape function has been lately used to model due to the good and easy fitting with few parameters (mainly from three to six parameters). The least square method (LSM) and maximum likelihood function can be used to determine the logistic function's parameters (MLF) [34]. The logistic function is nonlinear and comprises four variables, making it a difficult problem to solve. To filter out noise Gaussian Kernel Optimization is used [35].

The model Support Vector Regression (SVR) performs well when compared to other model (LSTM, CNN, GRU). Support Vector Regression has the lowest Mean Absolute Error, Root Mean Squared Error when compared to other models. In order to optimize the model(Support Vector Regression) here two optimization algorithms are used. First one is the Artificial Bee Colony algorithm and the other is the Chaotic Particle Swarm Optimization algorithm. Though Particle Swarm Optimization trapped in the local optimum and it failed to achieve Global Optimization in order to achieve global optimization, Particle Swarm Optimization is enhanced with Chaotic Optimization and hence it can achieve Global Optimization. In Chaotic Optimization performance can be enhanced with logistic maps. The results after optimizing the model with CPSO optimisation is are shown in Fig. 14.

3.3.5.1 Root mean square error

The root mean square error (RMSE) is a commonly used metric for determining how much a value predicted by an estimator or model differs from the actual values that were observed.

3.3.5.2 Mean absolute error

A model evaluation statistic used with regression models is known as mean absolute error. The average of the absolute values of each prediction error over all test set instances is the mean absolute error of a model with respect to the test set.

3.3.5.3 Mean absolute percentage error and mean squared error

Mean Absolute Percentage Error is the most widely used measure for checking forecast accuracy. It belongs to the category of scale-independent percentage errors, which can be used to compare series on various scales. The Mean Squared Error (MSE) is a measure of how close a fitted line is to data points.

From the Table 3 it is inferred that the result SVR with Chaotic Particle Optimization outperforms the SVR with ABC Optimization.

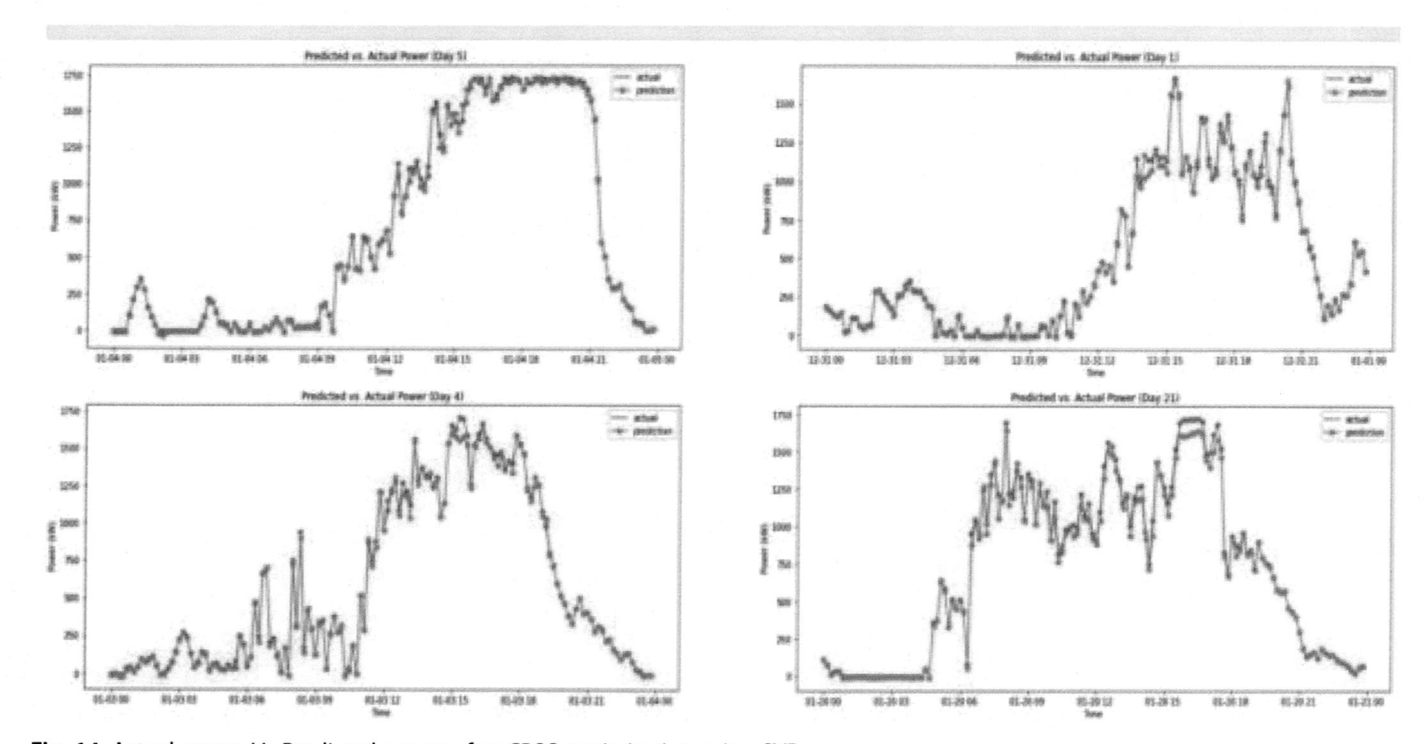

Fig. 14 Actual power Vs Predicted power after CPSO optimisation using SVR.

Table 3 Performance Analysis of SVR with ABC and CPSO optimization algorithms.

Model	MAE	RMSE
SVR with Chaotic Particle Swarm Optimisation	20.22	50.26
SVR with ABC Optimization	39.70	68.4

4. Conclusion and future works

Wind power prediction based on deep learning approaches are formulated and discussed. The literature comparison of different approaches and their advantages and downsides of various related methods as well as their improvement strategies are discussed. Here the noise is detected using Gaussian kernel and effectively filtered using Adaptive filtering method. Three different feature selection algorithms i.e., Decision tree, Linear regression, Random forest have been employed. Among these algorithms, Decision tree outperforms other two algorithms. For wind power prediction different models i.e., Support Vector Regressor(SVR),Convolutional Neural Network(CNN),Gated Recurrent Unit(GRU), Long-Short Term Memory(LSTM) have been implemented. On comparing with other models, SVR performs better than other models. Finally SVR is optimized using Chaotic Particle Swarm Optimisation (CPSO) to predict wind power more accurately. The optimized SVR model achieves better results. The predicted wind power measurements is close to the actual measurements.

In the future, the problem of missing data can be solved using different imputation methods. The prediction model can be further extended to predict other parameters like faults in the wind turbine. In this paper, Chaotic Particle Swarm Optimization is implemented to achieve global optimum. In future, different optimization techniques with different models can be used to achieve greater results. The knowledge gained on the influence of temperature, humidity and air pressure on the error of a neural network based model for wind power prediction can be expanded to different types of models. It would be of interest to see how this influence compares to other types of models.

Acknowledgment

We would like to thank Professor and our Research Supervisor Dr. P. Anandha Kumar, for their patient instructions, passionate support, and constructive criticisms of this study effort. We also want to thank Mr. Alex for his guidance and help in keeping our progress on schedule. We are also grateful for the insightful comments offered by our Professor Dr. Dhananjay Kumar.

Funding statement

We received no specific funding for this study.

Conflicts of Interest

We have no conflicts of interest to report regarding this study.

References

[1] K. Yuan, K. Zhang, Y. Zheng, D. Li, Y. Wang, Z. Yang, Irregular distribution of wind power prediction, J. Mod. Power Syst. Clean Energy 6 (6) (2018) 1172–1180.
[2] X. Liu, L. Yang, Z. Zhang, Short-term multi-step ahead wind power predictions based on a novel deep convolutional recurrent network method, IEEE Trans. Sustain. Energy 12 (3) (2021) 1820–1833.
[3] J. Nielson, K. Bhaganagar, R. Meka, A. Alaeddini, Using atmosp.116273 pheric inputs for artificial neural networks to improve wind turbine power prediction, Energy 190 (2020).
[4] G. An, Z. Jiang, X. Cao, Y. Liang, Y. Zhao, Z. Li, W. Dong, H. Sun, Short-term wind power prediction based on particle swarm optimization-extreme learning machine model combined with Adaboost algorithm, IEEE Access 9 (2021) 94040–94052.
[5] Y. Wang, P. Song, H. Liu, L. Wu, A probabilistic wind power forecasting approach based on gaussian process regression, in: 2020 IEEE/IAS Industrial and Commercial Power System Asia (I&CPS Asia), 2020.
[6] Z. Cao, Y. Zhang, J. Guan, S. Zhou, G. Wen, A chaotic ant colony optimized link prediction algorithm, IEEE Trans. Syst. Man Cybern.: Syst. 51 (9) (2021) 5274–5288.
[7] A.S. Qureshi, A. Khan, A. Zameer, A. Usman, Wind power prediction using deep neural network based meta regression and transfer learning, Appl. Soft Comput. 58 (2017) 742–755.
[8] R.P. Shetty, A. Sathyabhama, P.S. Pai, Comparison of modeling methods for wind power prediction: a critical study, Front. Energy 14 (2) (2018) 347–358.
[9] X. Meng, R. Wang, X. Zhang, M. Wang, H. Ma, Z. Wang, Hybrid neural network based on GRU with uncertain factors for forecasting ultra-short-term wind power, in: 2020 2nd International Conference on Industrial Artificial Intelligence (IAI), 2020.
[10] R. Fang, Y. Wang, R. Shang, Y. Liang, L. Wang, C. Peng, The ultra-short term power prediction of wind farm considering operational condition of wind turbines, Int. J. Hydrogen Energy 41 (35) (2016) 15733–15739.
[11] Y. Mousavi, A. Alfi, I.B. Kucukdemiral, Enhanced fractional chaotic whale optimization algorithm for parameter identification of isolated wind-diesel power systems, IEEE Access 8 (2020) 140862–140875.
[12] E. Taslimi Renani, M.F.M. Elias, N.A. Rahim, Using data-driven approach for wind power prediction: a comparative study, Energ. Conver. Manage. 118 (2016) 193–203.
[13] Z. Lin, X. Liu, M. Collu, Wind power prediction based on high-frequency SCADA data along with isolation forest and deep learning neural networks, Int. J. Electr. Power Energy Syst. 118 (2020) 105835.
[14] J. Yan, T. Ouyang, Advanced wind power prediction based on data-driven error correction, Energ. Conver. Manage. 180 (2019) 302–311.
[15] R. Donida Labati, A. Genovese, V. Piuri, F. Scotti, G. Sforza, A decision support system for wind power production, IEEE Trans. Syst. Man Cybern.: Syst. 50 (1) (2020) 290–304.
[16] O. Eyecioglu, B. Hangun, K. Kayisli, M. Yesilbudak, Performance comparison of different machine learning algorithms on the prediction of wind turbine power generation, in: 2019 8th International Conference on Renewable Energy Research and Applications (ICRERA), 2019.

[17] S. Daneshvar Dehnavi, A. Shirani, H. Mehrjerdi, M. Baziar, L. Chen, New deep learning-based approach for wind turbine output power modeling and forecasting, IEEE Trans. Ind. Appl. (2020) 101–113.
[18] F. Harrou, A. Saidi, Y. Sun, Wind power prediction using bootstrap aggregating trees approach to enabling sustainable wind power integration in a smart grid, Energ. Conver. Manage. 201 (2019) 112077.
[19] J. Heinermann, O. Kramer, Machine learning ensembles for wind power prediction, Renew. Energy 89 (2016) 671–679.
[20] Q. Hu, S. Zhang, M. Yu, Z. Xie, Short-term wind speed or power forecasting with heteroscedastic support vector regression, IEEE Trans. Sustain. Energy 7 (1) (2016) 241–249.
[21] J. Lee, W. Wang, F. Harrou, Y. Sun, Wind power prediction using ensemble learning-based models, IEEE Access 8 (2020) 61517–61527.
[22] T. Ouyang, X. Zha, L. Qin, A combined multivariate model for wind power prediction, Energ. Conver. Manage. 144 (2017) 361–373.
[23] R. Pandit, A. Kolios, SCADA data-based support vector machine wind turbine power curve uncertainty estimation and its comparative studies, Appl. Sci. 10 (23) (2020) 8685.
[24] A. Tascikaraoglu, M. Uzunoglu, A review of combined approaches for prediction of short-term wind speed and power, Renew. Sustain. Energy Rev. 34 (2014) 243–254.
[25] P. Lu, L. Ye, Y. Zhao, B. Dai, M. Pei, Y. Tang, Review of meta-heuristic algorithms for wind power prediction: methodologies, applications and challenges, Appl. Energy 301 (2021) 117446.
[26] Q. Xiaoyun, K. Xiaoning, Z. Chao, J. Shuai, M. Xiuda, Short-term prediction of wind power based on deep long short-term memory, in: IEEE PES Asia-Pacific Power and Energy Engineering Conference (APPEEC), 2016.
[27] F. Shahid, A. Zameer, A. Mehmood, M.A.Z. Raja, A novel wavenets long short term memory paradigm for wind power prediction, Appl. Energy 269 (2020) 115098.
[28] B. Zhou, X. Ma, Y. Luo, D. Yang, Wind Power Prediction Based on LSTM Networks and Nonparametric Kernel Density Estimation, IEEE Access 7 (2019) 165279–165292.
[29] C. Li, G. Tang, X. Xue, A. Saeed, X. Hu, Short-term wind speed interval prediction based on ensemble GRU model, IEEE Trans. Sustain. Energy 11 (3) (2020) 1370–1380.
[30] H. Leng, X. Li, J. Zhu, H. Tang, Z. Zhang, N. Ghadimi, A new wind power prediction method based on ridgelet transforms, hybrid feature selection and closed-loop forecasting, Adv. Eng. Inform. 36 (2018) 20–30.
[31] Y. Li, F. Yang, W. Zha, L. Yan, Combined optimization prediction model of regional wind power based on convolution neural network and similar days machines, 8 (4) (2020) 80.
[32] Z. Xu, W. Yixian, C. Yunlong, C. Xueting, G. Lei, Short-term wind speed prediction based on GRU, in: 2019 IEEE Sustainable Power and Energy Conference (iSPEC), 2019.
[33] Y. Zhang, H. Sun, Y. Guo, Wind power prediction based on PSO-SVR and Grey combination model, IEEE Access 7 (2019) 136254–136267.
[34] J. Yan, H. Zhang, Y. Liu, S. Han, L. Li, Uncertainty estimation for wind energy conversion by probabilistic wind turbine power curve modelling, Appl. Energy 239 (2019) 1356–1370.
[35] A. Ahmadpour, S. Gholami Farkoush, Gaussian models for probabilistic and deterministic wind power prediction: wind farm and regional, Int. J. Hydrogen Energy 45 (51) (2020) 27779–27791.

About the authors

Preetha Evangeline David is currently working as an Associate Professor and Head of the Department in the Department of Artificial Intelligence and Machine Learning at Chennai Institute of Technology, Chennai, India. She holds a PhD from Anna University, Chennai in the area of Cloud Computing. She has published many research papers and Patents focusing on Artificial Intelligence, Digital Twin Technology, High Performance Computing, Computational Intelligence and Data Structures. She is currently working on Multi-disciplinary areas in collaboration with other technologies to solve socially relevant challenges and provide solutions to human problems.

Anandhakumar is a professor in the Department of Information Technology at Anna University, Chennai. He has completed his doctorate in the year 2006 from Anna University. He has produced 17 PhD's in the field of Image Processing, Cloud Computing, Multimedia Technology and Machine Learning. His ongoing research lies in the field of Digital Twin Technology, Machine Learning and Artificial Intelligence. He has published more than 150 papers indexed in SCI, SCOPUS, WOS, etc.

Cyber data trend and intelligent computing

Preetha Evangeline David[a] and Atma Sahu[b]
[a]Department of Artificial Intelligence and Machine Learning, Chennai Institute of Technology, Chennai, Tamil Nadu, India
[b]Coppin State University, Baltimore, MD, United States

Contents

Advances in Computers, Volume 132
ISSN 0065-2458
https://doi.org/10.1016/bs.adcom.2023.08.005

Abstract

A zero trust cyber security posture provides the opportunity to create a more robust and resilient security, simplifies security management, improve end-user experience, and enable modern IT practices. Conventional castle-and-moat cyber security models, which rely on secure network perimeters and virtual private network-based employee and third-party remote access, are proving to be no match for evolving cyber threats, particularly as business models and workforce dynamics evolve. The anticipated growth of smart devices, 5G, edge computing, and artificial intelligence promises to create even more data, connected nodes, and expanded attack surfaces. Despite making significant investments in security organizations continue to struggle with security breaches. It is highly challenging to analyze the data that is flowing into the security operation centers. This doesn't include the information feeds from network devices, application data, and other inputs across the broader technology stack that are often targets of advanced attackers looking for new vectors or using new malware. Meanwhile, the cost of cybercrime continues to climb. It's time to call for AI backup. Cyber AI can be a force multiplier that enables organizations not only to respond faster than attackers can move, but also to anticipate these moves and react to them in advance.

1. Introduction

AI's ability to adaptively learn and detect novel patterns can accelerate detection, containment, and response, easing the burden on SOC analysts and allowing them to be more proactive. It can help organizations prepare for the eventual development of AI-driven cybercrimes. Increase in network-connected devices. 5G, IoT, Wi-Fi 6, and other networking advances are driving an increase in network-connected devices. When seeking a soft attack vector, cybercriminals will be able to choose from a growing number of networks. The unprecedented number of devices connected to these networks produces data that needs to be processed and secured, contributing to the data logjam in the SOC. It can be challenging to keep track of and manage active assets, their purpose, and their expected behavior, especially when they're managed by service orchestrators.

Rather than being centrally located and controlled, many of these devices are spread across various remote locations, operating in multiple edge environments where they collect data to send back to the enterprise. Without proper security precautions, devices can be compromised and continue to appear to operate normally on the network, essentially becoming intruder-controlled bots that can release malicious code or conduct swarm-based attacks.

An increasingly global supply chain and hosted data, infrastructure, and services have long contributed to third-party risk. And as more and more organizations integrate data with third-party applications, APIs are a growing security concern. Gartner predicts that by 2022, API abuses will become the enterprise's most frequent attack vector. Third-party breaches are growing in complexity. Five years ago, an intruder might use widely available malware to target specific computer systems, gain contractor credentials and steal customer data—messy, to be sure, but with a clear source and the ability to monitor and remediate the damage. Such an attack pales in comparison to today's sophisticated intrusions, in which information stolen from one company can be used to compromise thousands of its customers and suppliers. Supply chain attacks can do the same by exploiting the least-secure embedded components of complex supply networks.

A breach with no boundaries can be nearly impossible to monitor and remediate, with active theft potentially continuing for many years. 5G is expected to completely transform enterprise networks with new connections, capabilities, and services. But the shift to 5G's mix of hardware- and distributed, software-defined networks, open architectures, and virtualized infrastructure will create new vulnerabilities and a larger attack surface, which will require more dynamic cyber protection. As public 5G networks expand, organizations in government, automotive, manufacturing, mining, energy, and other sectors have also begun to invest in private 5G networks that meet enterprise requirements for lower latency, data privacy, and secure wireless connectivity. From autonomous vehicles and drones to smart factory devices and mobile phones, an entire ecosystem of public and private 5G network–connected devices, applications, and services will create additional potential entry points for hackers. Each asset will need to be configured to meet specific security requirements. And with the increasing variety of devices, the network becomes more heterogeneous and more challenging to monitor and protect.

2. AI defense against today's cyber threats

Expanding attack surfaces and the escalating severity and complexity of cyber threats are exacerbated by a chronic shortage of cyber security talent. Employment in the field would have to grow by approximately 89% to eliminate the estimated global shortage of more than 3 million cyber security professionals.AI can help fill this gap.

Accelerated threat detection: Threat detection was one of the earliest applications of cyber AI. It can augment existing attack surface management techniques to reduce noise and allow scarce security professionals to zero in on the strongest signals and indicators of compromise. It can also make decisions and take action more rapidly and focus on more strategic activities. Advanced analytics and machine learning platforms can quickly sift through the high volume of data generated by security tools, identify deviations from the norm, evaluate the data from the thousands of new connected assets that are flooding the network, and be trained to distinguish between legitimate and malicious files, connections, devices, and users.

AI-driven network and asset mapping and visualization platforms can provide a real-time understanding of an expanding enterprise attack surface. They can identify and categorize active assets, including containerized assets, which can provide visibility into rogue asset behavior. Supply chain risk management software incorporating AI and machine learning can automate the processes of monitoring physical and digital supply chain environments and tracking the way assets are composed and linked.

AI can also serve as a force multiplier that helps security teams automate time-consuming activities and streamline containment and response. Consider machine learning, deep learning, natural language processing, reinforcement learning, knowledge representation, and other AI approaches. When paired with automated evaluation and decision-making, AI can help analysts manage an escalating number of increasingly complex security threats and achieve scale. For example, like its predecessors, 5G is vulnerable to jamming attacks, in which attackers deliberately interfere with signal transfer. By implementing an AI- based interference scheme and machine learning models, a real-time vulnerability assessment system was developed that could detect the presence of low-level signal interference and classify jamming patterns.

Automation can help maximize AI's impact and shrink the time between detection and remediation. SOC automation platforms embedded with AI and machine learning can take autonomous, preventative action, for example, blocking access to certain data—and escalate issues to the SOC for further evaluation. When layered on top of the API management solutions that control API access, machine learning models trained on user access patterns can inspect all API traffic to uncover, report on, and act on anomalies in real time.

Proactive security posture: Properly trained AI can enable a more proactive security posture and promote cyber resilience, allowing organizations to stay in operation even when under attack and reducing the amount of time an adversary is in the environment. For example, context-rich user behavior analytics can be combined with unsupervised machine learning algorithms to automatically examine user activities; recognize typical patterns in network activity or data access; identify, evaluate, and flag anomalies (and disregard false alarms); and decide if response or intervention is warranted. And by feeding intelligence to human security specialists and enabling them to actively engage in adversary pursuit, AI enables proactive threat hunting.

3. Understanding cyber security data

Data science is largely driven by the availability of data. Datasets typically represent a collection of information records that consist of several attributes or features and related facts, in which cyber security data science is based on. Thus, it's important to understand the nature of cyber security data containing various types of cyber-attacks and relevant features. The reason is that raw security data collected from relevant cyber sources can be used to analyze the various patterns of security incidents or malicious behavior, to build a data-driven security model to achieve our goal. Several datasets exist in the area of cyber security including intrusion analysis, malware analysis, anomaly, fraud, or spam analyses that are used for various purposes. Effectively analyzing and processing of these security features, building target machine learning-based security model according to the requirements, and eventually, data-driven decision making, could play a role to provide intelligent cyber security services. In Table 1, we summarize several such datasets including their various features and attacks that are accessible on the Internet, and highlight their usage based on machine learning techniques in different cyber applications.

4. AI-driven cybercrimes

The same features that make AI a valuable weapon against security threats—speedy data analysis, event processing, anomaly detection, continuous learning, and predictive intelligence—can also be manipulated by criminals to develop new or more effective attacks and detect system weaknesses. For example, researchers have used generative adversarial networks—two neural networks that compete against each other to create datasets similar to training data—to successfully crack millions of passwords.

Table 1 A summary of cyber security datasets highlighting diverse attack-types and machine learning-based usage in different cyber applications.

Dataset	Description
ADFA IDS	An intrusion dataset with different versions named ADFA-LD and ADFA-WD issued by the Australian Defense Academy (ADFA). They are designed for evaluation by host-based IDS
CERT	The dataset includes users activity logs that were created for the purpose of validating insider-threat detection systems. This can be used to analyze ML based user behavioral activities
DARPA	Intrusion detection dataset that includes LLDOS 1.0 and LLDOS 2.0.2 attack scenario data. Data traffic and attacks containing in DARPA are collected by MIT Lincoln Laboratory for evaluating network intrusion detection systems
KDD'99 Cup	Most widely used data set containing 41 features for evaluating anomaly detection methods, where attacks are categorized into four major classes, such as denial of service (DoS), remote-to-local (R2L), user-to-remote (U2R), and probing. KDD'99 Cup dataset can be used to evaluate ML-based attack detection model
CAIDA	The datasets CAIDA'07 and CAIDA'08 contain DDoS attack traffic and normal traffic traces. Thus CAIDA DDoS dataset can be used to evaluate ML-based DDoS attack detection model and inferring Internet Denial-of-Service activity
ISCX'12	The dataset contains 19 features and 19.11% of the traffic belongs to DDoS attacks. ISCX'12 was produced at the Canadian Institute for Cyber security and can be used to evaluate the effectiveness of ML-based network intrusion detection modeling
CIC-DDoS2019	The datasets include different attack scenarios, namely Brute-force, Heart bleed, Botnet, HTTP DoS, DDoS, Web attacks, and insider attack, collected by the Canadian Institute for Cyber security. Datasets can be used for evaluating ML based intrusion detection systems including Zero-Day attacks
MAWI	A collection of Japanese network research institutions and academic institutions used to detect and evaluate DDoS intrusions using ML technique
Bot-IoT	A dataset that incorporates legitimate and simulated IoT network traffic, along with different attacks for network forensic analytics in the area of Internet of Things. Bot-IoT can be used to evaluate the reliability using different statistical and machine learning methods for forensics purposes

Table 1 A summary of cyber security datasets highlighting diverse attack-types and machine learning-based usage in different cyber applications.—cont'd

Dataset	Description
DGA	The Alexa Top Sites dataset is generally used as a source of benign domain names. The malicious domain names are obtained from OSINT and DGArchive. DGA dataset can be used for experiments in ML-based automatic DGA domains classification or botnet detection.
CTU-13	A labeled malware dataset including botnet, normal, and background traffic that was captured at CTU University, Czech Republic. CTU-13 can be used for data-driven malware analysis using ML techniques and to evaluate the malware detection system

Similarly, an open-source, deep learning language model known as GPT-3 can learn the nuances of behavior and language. It could be used by cybercriminals to impersonate trusted users and make it nearly impossible to distinguish between genuine and fraudulent email and other communications. Phishing attacks could become far more contextual and believable.

Advanced adversaries can already infiltrate a network and maintain a long-term presence without being detected, typically moving slowly and discreetly, with specific targets. Add AI malware to the mix, and these intruders could learn how to quickly disguise themselves and evade detection while compromising many users and rapidly identifying valuable datasets. Organizations can help prevent such intrusions by fighting fire with fire: With enough data, AI-driven security tools can effectively anticipate and counter AI-driven threats in real time.

For example, security pros could leverage the same technique that researchers used to crack passwords to measure password strength or generate decoy passwords to help detect breaches. And contextual machine learning can be used to understand email users' behaviors, relationships, and time patterns to dynamically detect abnormal or risky user behavior. Approaches such as machine learning, natural language processing, and neural networks can help security analysts distinguish signal from noise. Using pattern recognition, supervised and unsupervised machine learning algorithms, and predictive and behavioral analytics, AI can help identify and repel attacks and automatically detect abnormal user behavior, allocation of network resources, or other anomalies. AI can be used to secure both on-premises architecture and enterprise cloud services, although securing workloads and resources in the cloud is typically less challenging than in legacy on-premises environments. On its own, AI (or any other technology, for that matter) isn't going to solve today's or tomorrow's complex security challenges. AI's ability to identify patterns and adaptively learn in real time as events warrant can accelerate detection, containment, and response; help reduce the heavy load on SOC analysts; and enable them to be more proactive. These workers will likely remain in high demand, but AI will change their roles. Organizations likely will need to reskill and retrain analysts to help change their focus from triaging alerts and other lower-level skills to more strategic, proactive activities. Finally, as the elements of AI- and machine learning–driven security threats begin to emerge, AI can help security teams prepare for the eventual development of AI-driven cybercrimes.

The intersection of AI and cyber security has been talked about for nearly a decade. Until now, those conversations revolved around buzzwords and rule-based products. Thanks to advances in computing power and storage capacity, we now see cyber security vendors starting to truly incorporate machine learning and AI into their products. Today, large enterprises can rely on such vendors to advance threat intelligence. Premier cyber security vendors have deployments across many enterprises, which serve as sensors for picking up data. By applying AI to the anonymized data from each customer, vendors can use the threat data from one organization to look for signs of similar breaches elsewhere. The network effects can be exponential: The bigger and more diverse the dataset, the more these vendors' detection improves and the greater their protection. For this reason, medium and large enterprises alike could benefit from working with managed service providers. Or, alternately, they can have their data science and cyber security teams work together to train AI models in their own cyber security warehouses. Today's computing power allows the development of sophisticated user and entity behavior analytics that detect signatures of bad actors or deviations from normal behavior.

5. Executive perspectives

5.1 Strategy

Cyber risk is a more important strategic concern than ever. With the amount of data organizations collect and the breadth of their partnerships and workforce, protection is growing more complicated. Cyber AI is now a leading practice for guarding against the volume and sophistication of recent cyberattacks. CEOs should be asking questions to understand the current security posture, and whether it needs to be upgraded. By positioning AI as a security and strategy priority, leaders can help their organizations align on the importance of strengthening defenses and managing risk.

5.2 Finance

As the prevalence and financial impact of cyber-attacks increase, the role in overseeing risks management. They should use their unique role in C-suite leadership to advocate for a fully funded enterprise wide adoption of AI-enhanced cyber defense. They can work with their cyber security teams to understand the investment required, timeline, risks, and benefits of cyber AI, and then present that information to the board as a key priority.

5.3 Risk

Bad actors have been leveraging AI for years to conduct cyber-attacks. CROs should prepare their organizations for the new normal of fighting those attacks with AI defense and intelligent security operations. Organizations should find internal support to build these new capabilities or evaluate outsourcing cyber protection to augment their security teams. Of course, AI defenses have their own vulnerabilities, and the threat landscape will continue to evolve. Acting now to begin improving defenses gradually rather than reacting when it's too late can help organizations protect customers and their data.

6. Zero trust

With cloud now mainstream, businesses managing services across multiple cloud providers are responsible for securing these technologies. As an enterprise more frequently relies on third-party vendors to host and manage data, infrastructure, and other services, the attack surface expands. Zero trust architecture is more risk-driven and context-aware, it recognizes the inconsistency, automatically denies the access request, and raises an alert. Automated response capabilities could be triggered to temporarily disable the user's account, given the likelihood that its credentials have been compromised.

Proper design and engineering of zero trust architectures can result in simple, modular environments and straightforward user access control and management. Streamlining the security stack can eliminate considerable management headaches, significantly reduce operational overhead, and help scale to 10 of 1000 of users. Similarly, onboarding employees, contractors, cloud service providers, and other vendors can become more efficient, flexible, responsive and secure.

The use of APIs across the technology ecosystem can facilitate system management in a zero trust manner by providing a consistent control layer. And cloud-based services enable organizations to leverage the substantial ongoing security investments of cloud vendors.

A final key element of the zero trust approach is micro segmenting networks, data, applications, workloads, and other resources into individual, manageable units to contain breaches and wrap security controls at the lowest level possible. By limiting access based on the principle of least privilege, a minimum set of users, applications, and devices has access to data and applications.

By removing the assumption of trust from the security architecture and authenticating every action, user, and device, zero trust helps enable a more robust and resilient security posture. The organizational benefits are complemented by a considerable end-user perk: seamless access to the tools and data needed to work efficiently.

7. Beefing up basic cyber hygiene

The zero trust mindset shift brings with it a set of design principles that guide security architecture development and build on existing security investments and processes. To enforce access control, companies must have situational awareness of their data and assets; companies that lag on basic cyber hygiene principles and practices may be challenged to realize the full benefits of zero trust. Fundamentals include:

7.1 Data discovery and classification

Data governance, inventory, classification, and tagging are critical. To create the appropriate trust zones and access controls, organizations need to understand their data, the criticality of that data, where it resides, how it is classified and tagged, and the people and applications that should have access to it.

7.2 Asset discovery and attack-surface management

Many organizations lack a real-time, updated inventory of all IT resources—including cloud resources, IP addresses, subdomains, application mapping, code repositories, social media accounts, and other external or internet-facing assets—and therefore can't identify security issues across the complete attack surface. To facilitate risk-based policy decisions surrounding their assets, it's critical for organizations to understand the enterprise IT environment.

7.3 Configuration and patch management

Without the ability to efficiently manage and document baseline configurations of key technology systems, deploy appropriate patches, test patched systems, and document new configurations, companies cannot easily identify changes and control risks to these systems. Malicious actors can exploit any vulnerability to gain a foothold within an organization.

7.4 Identity and access management

To ensure that access to technology resources is granted to the proper people, devices, and other assets, enterprises need to standardize and automate their identity life cycle management processes. They can extend their operations beyond traditional boundaries while protecting critical resources and maintaining an efficient user experience by moving the identity stack to the cloud, consuming identity-as-a-service, or implementing such advanced authentication methods as physical biometrics, behavioral monitoring, and conditional access.

7.5 Third-party risk management

To fully understand their entire risk surface, organizations need greater visibility into cyber risks related to their supply chains and ecosystem partners, including suppliers to third-party vendors.

7.6 Logging and monitoring

To identify potentially malicious incidents and issues, security teams need automated logging and monitoring systems with advanced AI and machine learning capabilities to help simplify the process of tracking, analyzing, and correlating data from volumes of detailed logs as well as alerts generated by internal and external systems, security controls, networks, and processes.

8. Engineered security automation and orchestration

Many security operations center (SOC) teams are challenged to keep pace with the volume of information generated by their technology and security controls. They must monitor, manage, and act upon continuous alerts and streams of data generated by fragmented security architectures and disparate, disconnected tools.

The number and nature of risk factors interrogated to support zero trust authentication and authorization—users, devices, or credentials and contextual data points such as location, privileges, application requirements, and behaviors warrants a more automated approach to monitoring, decision-making, enforcement, and auditing.

Many existing security technologies can be leveraged to build out zero trust architectures. To ensure more efficient automation and orchestration, zero trust adopters should rationalize the security stack and eliminate unnecessary and duplicative technologies or those that contribute to data overload, delay detection and response, and complicate system maintenance and management.

With a simplified security stack, existing systems and tools can be integrated via API connections to a security orchestration, automation, and response (SOAR) platform that can automate workflows and repetitive and manual tasks, and coordinate the flow of data and alerts to the SOC. SOAR platforms help add context to triggered events and can auto-remediate identified and known vulnerabilities, enabling staff to keep pace with incoming alerts and notifications, improving operational efficiency and accuracy, and decreasing response time.

"Migrating to zero trust architectures can seem like a heavy lift, especially in large enterprises saddled with legacy technologies and a lot of technical debt," says a senior technology leader at a large financial institution. "You have to break it into manageable chunks where you can identify a discrete win, such as deploying pervasive endpoint segmentation, and understand that win as part of your larger story of operationalizing zero trust."

9. Machine learning tasks in cyber security

Machine learning (ML) is typically considered as a branch of "Artificial Intelligence," which is closely related to computational statistics, data mining and analytics, data science, particularly focusing on making the computers to learn from data. Thus, machine learning models typically comprise of a set of rules, methods, or complex "transfer functions" that can be applied to find interesting data patterns, or to recognize or predict behavior, which could play an important role in the area of cyber security. In the following, we discuss different methods that can be used to solve machine learning tasks and how they are related to cyber security tasks.

9.1 Supervised learning

Supervised learning is performed when specific targets are defined to reach from a certain set of inputs, i.e., task-driven approach. In the area of machine learning, the most popular supervised learning techniques are known as classification and regression methods. These techniques are popular to classify or predict the future for a particular security problem. For instance, to predict denial-of-service attack (yes, no) or to identify different classes of network attacks such as scanning and spoofing, classification techniques can be used in the cyber security domain. ZeroR, OneR, Navies Bayes, Decision Tree, K-nearest neighbors, support vector machines, adaptive boosting, and logistic regression are the well-known classification techniques. In addition, recently Sarker et al. have proposed BehavDT, and IntruDtree classification techniques that are able to effectively build a data-driven predictive model. On the other hand, to predict the continuous or numeric value, e.g., total phishing attacks in a certain period or predicting the network packet parameters, regression techniques are useful. Regression analyses can also be used to detect the root causes of cybercrime and other types of fraud. Linear regression, support vector regression are the popular regression techniques. The main difference between classification and regression is that the output variable in the regression is numerical or continuous, while the predicted output for classification is categorical or discrete. Ensemble learning is an extension of supervised learning while mixing different simple models, e.g., Random Forest learning that generates multiple decision trees to solve a particular security task.

9.2 Unsupervised learning

In unsupervised learning problems, the main task is to find patterns, structures, or knowledge in unlabeled data, i.e., data-driven approach. In the area of cyber security, cyber-attacks like malware stays hidden in some ways, include changing their behavior dynamically and autonomously to avoid detection. Clustering techniques, a type of unsupervised learning, can help to uncover the hidden patterns and structures from the datasets, to identify indicators of such sophisticated attacks. Similarly, in identifying anomalies, policy violations, detecting, and eliminating noisy instances in data, clustering techniques can be useful. K-means, K-medoids are the popular partitioning clustering algorithms, and single linkage or complete linkage are the well-known hierarchical clustering algorithms used in various application domains. Moreover, a bottom-up clustering approach proposed by Sarker et al. can also be used by taking into account the data characteristics. Besides, feature engineering tasks like optimal feature selection or extraction related to a particular security problem could be useful for further analysis. Recently, Sarker et al. have proposed an approach for selecting security features according to their importance score values. Moreover, Principal component analysis, linear discriminant analysis, Pearson correlation analysis, or non-negative matrix factorization are the popular dimensionality reduction techniques to solve such issues.

Association rule learning is another example, where machine learning based policy rules can prevent cyber-attacks. In an expert system, the rules are usually manually defined by a knowledge engineer working in collaboration with a domain expert. Association rule learning on the contrary, is the discovery of rules or relationships among a set of available security features or attributes in a given dataset. To quantify the strength of relationships, correlation analysis can be used. Many association rule mining algorithms have been proposed in the area of machine learning and data mining literature, such as logic-based, frequent pattern based, tree-based, etc. Recently, Sarker et al. have proposed an association rule learning approach considering non-redundant generation that can be used to discover a set of useful security policy rules. Moreover, AIS, Apriori, Apriori-TID and Apriori-Hybrid, FP-Tree, and RARM and Eclat are the well-known association rule learning algorithms that are capable to solve such problems by generating a set of policy rules in the domain of cyber security.

9.3 Neural networks and deep learning

Deep learning is a part of machine learning in the area of artificial intelligence, which is a computational model that is inspired by the biological neural networks in the human brain. Artificial Neural Network (ANN) is frequently used in deep learning and the most popular neural network algorithm is back propagation. It performs learning on a multi-layer feed-forward neural network consists of an input layer, one or more hidden layers, and an output layer. The main difference between deep learning and classical machine learning is its performance on the amount of security data increases. Typically deep learning algorithms perform well when the data volumes are large, whereas machine learning algorithms perform comparatively better on small datasets. In our earlier work, Sarker et al. we have illustrated the effectiveness of these approaches considering contextual datasets.

However, deep learning approaches mimic the human brain mechanism to interpret large amount of data or the complex data such as images, sounds and texts. In terms of feature extraction to build models, deep learning reduces the effort of designing a feature extractor for each problem than the classical machine learning techniques. Beside these characteristics, deep learning typically takes a long time to train an algorithm than a machine learning algorithm, however, the test time is exactly the opposite. Thus, deep learning relies more on high-performance machines with GPUs than classical machine-learning algorithms. The most popular deep neural network learning models include multi-layer perceptron (MLP), convolutional neural network (CNN), recurrent neural network (RNN) or long-short term memory (LSTM) network. In recent days, researchers use these deep learning techniques for different purposes such as detecting network intrusions, malware traffic detection and classification, etc. in the domain of cyber security.

9.4 Other learning techniques

Semi-supervised learning can be described as a hybridization of supervised and unsupervised techniques discussed above, as it works on both the labeled and unlabeled data. In the area of cyber security, it could be useful, when it requires labeling data automatically without human intervention, to improve the performance of cyber security models. Reinforcement techniques are another type of machine learning that characterizes an agent by creating its own learning experiences through interacting directly with the environment, i.e., environment-driven approach, where the environment is typically formulated as a Markov decision process and take decision based on a reward function. Monte Carlo learning, Q-learning, Deep Q Networks, are the most common reinforcement learning algorithms. For instance, in a recent work, the authors present an approach for detecting botnet traffic or malicious cyber activities using reinforcement learning combining with neural network classifier.

In another work, the authors discuss about the application of deep reinforcement learning to intrusion detection for supervised problems, where they received the best results for the Deep Q-Network algorithm. In the context of cyber security, genetic algorithms that use fitness, selection, crossover, and mutation for finding optimization, could also be used to solve a similar class of learning problem. Various types of machine learning techniques discussed above can be useful in the domain of cyber security, to build an effective security model. In Table 2, we have summarized several machine learning techniques that are used to build various types of security models for various purposes. Although these models typically represent a learning-based security model, in this paper, we aim to focus on a comprehensive cyber security data science model and relevant issues, in order to build a data-driven intelligent security system. In the next section, we highlight several research issues and potential solutions in the area of cyber security data science.

Table 2 A summary of machine learning tasks in the domain of cyber security.

Technique	Purpose
SVM	To classify various attacks such as DoS, Probe, U2R, and R2L
SVM-PSO	To build intrusion detection system
FCM clustering, ANN and SVM	To build network intrusion detection system
KNN	Network intrusion detection system
Decision Tree	To detect the malicious code's behavior information by running malicious code on the virtual machine and analyze the behavior information for intrusion detection
Genetic Algorithm and Decision Tree	To solve the problem of small disjunct in the decision tree based intrusion detection system
Random Forests	To build network intrusion detection systems
Association Rule	To build network intrusion detection system
Behavior Rule	To build intrusion detection system for safety critical medical cyber physical systems
Deep Learning Recurrent, RNN, LSTM	To build anomaly intrusion detection system and attack classification
Deep and Reinforcement Learning	Malicious activities and intrusion detection system
Semi-supervised Adaboost	For network anomaly detection

10. Research issues and future directions

Our study opens several research issues and challenges in the area of cyber security data science to extract insight from relevant data towards data-driven intelligent decision making for cyber security solutions. In the following, we summarize these challenges ranging from data collection to decision making.

- Cyber security datasets: Source datasets are the primary component to work in the area of cyber security data science. Most of the existing datasets are old and might insufficient in terms of understanding the recent behavioral patterns of various cyber-attacks. Although the data can be transformed into a meaningful understanding level after performing several processing tasks, there is still a lack of understanding of the characteristics of recent attacks and their patterns of happening. Thus, further processing or machine learning algorithms may provide a low accuracy rate for making the target decisions. Therefore, establishing a large number of recent datasets for a particular problem domain like cyber risk prediction or intrusion detection is needed, which could be one of the major challenges in cyber security data science.
- Handling quality problems in cyber security datasets: The cyber datasets might be noisy, incomplete, insignificant, imbalanced, or may contain inconsistency instances related to a particular security incident. Such problems in a data set may affect the quality of the learning process and degrade the performance of the machine learning-based models. To make a data-driven intelligent decision for cyber security solutions, such problems in data is needed to deal effectively before building the cyber models. Therefore, understanding such problems in cyber data and effectively handling such problems using existing algorithms or newly proposed algorithm for a particular problem domain like malware analysis or intrusion detection and prevention is needed, which could be another research issue in cyber security data science.
- Security policy rule generation: Security policy rules reference security zones and enable a user to allow, restrict, and track traffic on the network based on the corresponding user or user group, and service, or the application. The policy rules including the general and more specific rules are compared against the incoming traffic in sequence during the execution, and the rule that matches the traffic is applied. The policy rules used in most of the cyber security systems are static and generated by human

expertise or ontology-based. Although, association rule learning techniques produce rules from data, however, there is a problem of redundancy generation that makes the policy rule-set complex. Therefore, understanding such problems in policy rule generation and effectively handling such problems using existing algorithms or newly proposed algorithm for a particular problem domain like access control is needed, which could be another research issue in cyber security data science.

- Hybrid learning method: Most commercial products in the cyber security domain contain signature-based intrusion detection techniques. However, missing features or insufficient profiling can cause these techniques to miss unknown attacks. In that case, anomaly-based detection techniques or hybrid technique combining signature-based and anomaly-based can be used to overcome such issues. A hybrid technique combining multiple learning techniques or a combination of deep learning and machine-learning methods can be used to extract the target insight for a particular problem domain like intrusion detection, malware analysis, access control, etc. and make the intelligent decision for corresponding cyber security solutions.
- Protecting the valuable security information: Another issue of a cyber-data attack is the loss of extremely valuable data and information, which could be damaging for an organization. With the use of encryption or highly complex signatures, one can stop others from probing into a dataset. In such cases, cyber security data science can be used to build a data-driven impenetrable protocol to protect such security information. To achieve this goal, cyber analysts can develop algorithms by analyzing the history of cyber-attacks to detect the most frequently targeted chunks of data. Thus, understanding such data protecting problems and designing corresponding algorithms to effectively handling these problems, could be another research issue in the area of cyber security data science.
- Context-awareness in cyber security: Existing cyber security work mainly originates from the relevant cyber data containing several low-level features. When data mining and machine learning techniques are applied to such datasets, a related pattern can be identified that describes it properly. However, a broader contextual information like temporal, spatial, relationship among events or connections, dependency can be used to decide whether there exists a suspicious activity or not. For instance, some approaches may consider individual connections as DoS attacks, while security experts might not treat them as malicious by themselves. Thus, a significant limitation of existing cyber security

work is the lack of using the contextual information for predicting risks or attacks. Therefore, context-aware adaptive cybersecurity solutions could be another research issue in cyber security data science.

- Feature engineering in cyber security: The efficiency and effectiveness of a machine learning-based security model has always been a major challenge due to the high volume of network data with a large number of traffic features. The large dimensionality of data has been addressed using several techniques such as principal component analysis (PCA), singular value decomposition (SVD), etc. In addition to low-level features in the datasets, the contextual relationships between suspicious activities might be relevant. Such contextual data can be stored in an ontology or taxonomy for further processing. Thus how to effectively select the optimal features or extract the significant features considering both the low-level features as well as the contextual features, for effective cyber security solutions could be another research issue in cyber security data science.
- Remarkable security alert generation and prioritizing: In many cases, the cyber security system may not be well defined and may cause a substantial number of false alarms that are unexpected in an intelligent system. For instance, an IDS deployed in a real-world network generates around nine million alerts per day. A network based intrusion detection system typically looks at the incoming traffic for matching the associated patterns to detect risks, threats or vulnerabilities and generate security alerts. However, to respond to each such alert might not be effective as it consumes relatively huge amounts of time and resources, and consequently may result in self-inflicted DoS. To overcome this problem, a high-level management is required that correlate the security alerts considering the current context and their logical relationship including their prioritization before reporting them to users, which could be another research issue in cyber security data science.
- Recent analysis in cyber security solutions: Machine learning-based security models typically use a large amount of static data to generate data-driven decisions. Anomaly detection systems rely on constructing such a model considering normal behavior and anomaly, according to their patterns. However, normal behavior in a large and dynamic security system is not well defined and it may change over time, which can be considered as an incremental growing of dataset. The patterns in incremental datasets might be changed in several cases. This often results in a substantial number of false alarms known as false positives. Thus, a recent

malicious behavioral pattern is more likely to be interesting and significant than older ones for predicting unknown attacks. Therefore, effectively using the concept of recent analysis in cyber security solutions could be another issue in cyber security data science.

11. Discussion

Although several research efforts have been directed towards cyber security solutions, discussed in "Background," "Cyber security data science," and "Machine learning tasks in cyber security" sections in different directions, this paper presents a comprehensive view of cyber security data science. For this, we have conducted a literature review to understand cyber security data, various defense strategies including intrusion detection techniques, different types of machine learning techniques in cyber security tasks. Based on our discussion on existing work, several research issues related to security datasets, data quality problems, policy rule generation, learning methods, data protection, feature engineering, security alert generation, recent analysis, etc. are identified that require further research attention in the domain of cyber security data science.

The scope of cyber security data science is broad. Several data-driven tasks such as intrusion detection and prevention, access control management, security policy generation, anomaly detection, spam filtering, fraud detection and prevention, various types of malware attack detection and defense strategies, etc. can be considered as the scope of cyber security data science. Such tasks based categorization could be helpful for security professionals including the researchers and practitioners who are interested in the domain-specific aspects of security systems. The output of cyber security data science can be used in many application areas such as Internet of things (IoT) security, network security, cloud security, mobile and web applications, and other relevant cyber areas.

Moreover, intelligent cyber security solutions are important for the banking industry, the healthcare sector, or the public sector, where data breaches typically occur. Besides, the data-driven security solutions could also be effective in AI-based blockchain technology, where AI works with huge volumes of security event data to extract the useful insights using machine learning techniques, and block-chain as a trusted platform to store such data. Although in this paper, we discuss cyber security data science focusing on examining raw security data to data-driven decision making for intelligent security solutions, it could also be related to big data analytics in terms of data processing and decision making. Big data deals with data sets that are too large or complex having characteristics of high data volume, velocity, and variety. Big data analytics mainly has two parts consisting of data management involving data storage, and analytics. The analytics typically describe the process of analyzing such datasets to discover patterns, unknown correlations, rules, and other useful insights. Thus, several advanced data analysis techniques such as AI, data mining, machine learning could play an important role in processing big data by converting big problems to small problems. To do this, the potential strategies like parallelization, divide-and-conquer, incremental learning, sampling, granular computing, feature or instance selection, can be used to make better decisions, reducing costs, or enabling more efficient processing. In such cases, the concept of cyber security data science, particularly machine learning-based modeling could be helpful for process automation and decision making for intelligent security solutions. Moreover, researchers could consider modified algorithms or models for handing big data on parallel computing platforms like Hadoop, Storm, etc. Based on the concept of cyber security data science discussed in the paper, building a data-driven security model for a particular security problem and relevant empirical evaluation to measure the effectiveness and efficiency of the model and to assess the usability in the real-world application domain could be a future work.

12. Conclusion

Motivated by the growing significance of cyber security and data science, and machine learning technologies, in this paper, we have discussed how cyber security data science applies to data-driven intelligent decision making in smart cyber security systems and services. We also have discussed how it can impact security data, both in terms of extracting insight of security incidents and the dataset itself. We aimed to work on cyber security data science by discussing the state of the art concerning security incidents data and corresponding security services. We also discussed how machine learning techniques can impact in the domain of cyber security, and examine the security challenges that remain. In terms of existing research, much focus has been provided on traditional security solutions, with less available work in machine learning technique based security systems. For each common technique, we have discussed relevant security research.

The purpose of this article is to share an overview of the conceptualization, understanding, modeling, and thinking about cyber security data science. We have further identified and discussed various key issues in security analysis to showcase the signpost of future research directions in the domain of cyber security data science. Based on the knowledge, we have also provided a generic multi-layered framework of cyber security data science model based on machine learning techniques, where the data is being gathered from diverse sources, and the analytics complement the latest data-driven patterns for providing intelligent security services. The framework consists of several main phases - security data collecting, data preparation, machine learning based security modeling, and incremental learning and dynamism for smart cyber security systems and services. We specifically focused on extracting insights from security data, from setting a research design with particular attention to concepts for data-driven intelligent security solutions.

Overall, this chapter aimed not only to discuss cyber security data science and relevant methods but also to discuss the applicability towards data-driven intelligent decision making in cyber security systems and services from machine learning perspectives. Our analysis and discussion can have several implications both for security researchers and practitioners. For researchers, we have highlighted several issues and directions for future research. Other areas for potential research include empirical evaluation of the suggested data-driven model, and comparative analysis with other security systems. For practitioners, the multi-layered machine learning-based model can be used as a reference in designing intelligent cyber security systems for organizations. We believe that our study on cyber security data science opens a promising path and can be used as a reference guide for both academia and industry for future research and applications in the area of cyber security.

Further reading

[1] Av-test Institute, Germany, https://www.av-test.org/en/statistics/malware/. Accessed 20 Oct 2019. 8. IBM Security Report, https://www.ibm.com/security/data-breach. Accessed on 20 Oct 2019.
[2] S. Mohammadi, H. Mirvaziri, M. Ghazizadeh-Ahsaee, H. Karimipour, Cyber intrusion detection by combined feature selection algorithm, J. Inform. Sec. Appl. 44 (2019) 80–88.
[3] Foroughi F, Luksch P.. Data Science Methodology for Cybersecurity Projects. arXiv preprint arXiv:1803.04219, 2018.
[4] K. Sigler, Crypto-jacking: how cyber-criminals are exploiting the crypto-currency boom, Comput. Fraud Sec. 2018 (9) (2018) 12–14.
[5] 2019 Data Breach Investigations Report. https://enterprise.verizon.com/resources/reports/dbir/. Accessed 20 Oct 2019.

[6] A. Khraisat, I. Gondal, P. Vamplew, J. Kamruzzaman, Survey of intrusion detection systems: techniques, datasets and challenges, Cybersecurity. 2 (1) (2019) 20.
[7] L. Johnson, Computer Incident Response and Forensics Team Management: Conducting a Successful Incident Response, Syngress, 2013. ISBN: 978-1597499965.
[8] I. Brahmi, H. Brahmi, S.B. Yahia, A multi-agents intrusion detection system using ontology and clustering techniques, in: IFIP International Conference on Computer Science and its Applications, Springer, New York, 2015, pp. 381–393.
[9] E. Viegas, A.O. Santin, A. Franca, R. Jasinski, V.A. Pedroni, L.S. Oliveira, Towards an energy-efficient anomaly-based intrusion detection engine for embedded systems, IEEE Trans. Comput. 66 (1) (2016) 163–177.
[10] Y. Xin, L. Kong, Z. Liu, Y. Chen, Y. Li, H. Zhu, M. Gao, H. Hou, C. Wang, Machine learning and deep learning methods for cybersecurity, IEEE Access. 6 (2018) 35365–35381.
[11] I. Dutt, S. Borah, I.K. Maitra, K. Bhowmik, A. Maity, S. Das, Real-time hybrid intrusion detection system using machine learning techniques, Advances in Communication, Devices and Networking, Springer, 2018, pp. 885–894.
[12] D.J. Ragsdale, C. Carver, J.W. Humphries, U.W. Pooch, Adaptation techniques for intrusion detection and intrusion response systems, in: SMC 2000 Conference Proceedings. 2000 IEEE International Conference on Systems, Man and Cybernetics.'Cybernetics Evolving to Systems, Humans, Organizations, and Their Complex Interactions'(cat. No. 0), vol. 4, IEEE, 2000, pp. 2344–2349.
[13] L. Cao, Data science: challenges and directions, Commun. ACM. 60 (8) (2017) 59–68.
[14] A. Rizk, A. Elragal, Data science: developing theoretical contributions in information systems via text analytics, J. Big Data. 7 (1) (2020) 1–26.
[15] R.P. Lippmann, D.J. Fried, I. Graf, J.W. Haines, K.R. Kendall, D. McClung, D. Weber, S.E. Webster, D. Wyschogrod, R.K. Cunningham, et al., Evaluating intrusion detection systems: the 1998 darpa of-line intrusion detection evaluation, in: Proceedings DARPA Information Survivability Conference and Exposition. DISCEX'00, vol. 2, IEEE, 2000, pp. 12–26.
[16] Kdd cup 99. http://kdd.ics.uci.edu/databases/kddcup99/kddcup99.html. Accessed 20 Oct 2019.
[17] Microsoft malware classification (big 2015). arXiv:org/abs/1802.10135/. Accessed 20 Oct 2019.
[18] N. Koroniotis, N. Moustafa, E. Sitnikova, B. Turnbull, Towards the development of realistic botnet dataset in the internet of things for network forensic analytics: bot-iot dataset, Future Gen. Comput. Syst. 100 (2019) 779–796.
[19] T.R. McIntosh, J. Jang-Jaccard, P.A. Watters, Large scale behavioral analysis of ransomware attacks, in: International Conference on Neural Information Processing, Springer, New York, 2018, pp. 217–229.
[20] J. Han, J. Pei, M. Kamber, Data Mining: Concepts and Techniques, in: The Morgan Kaufmann Series in Data Management Systems, 2011.
[21] I.H. Witten, E. Frank, Data Mining: Practical Machine Learning Tools and Techniques, in: The Morgan Kaufmann Series in Data Management Systems, 2005.
[22] S. Dua, X. Du, Data Mining and Machine Learning in Cyber Security, Auerbach Publications, 2016.
[23] M.V. Kotpalliwar, R. Wajgi, Classification of attacks using support vector machine (svm) on kddcup'99 ids database, in: 2015 Fifth International Conference on Communication Systems and Network Technologies, IEEE, 2015, pp. 987–990.
[24] M.S. Pervez, D.M. Farid, Feature selection and intrusion classification in nsl-kdd cup 99 dataset employing svms, in: The 8th International Conference on Software, Knowledge, Information Management and Applications (SKIMA 2014), IEEE, 2014, pp. 1–6.

[25] M. Yan, Z. Liu, A new method of transductive SVM-based network intrusion detection, in: International Conference on Computer and Computing Technologies in Agriculture, Springer, New York, 2010, pp. 87–95.
[26] Y. Li, J. Xia, S. Zhang, J. Yan, X. Ai, K. Dai, An efficient intrusion detection system based on support vector machines and gradually feature removal method, Expert. Syst. Appl. 39 (1) (2012) 424–430.
[27] M.G. Raman, N. Somu, S. Jagarapu, T. Manghnani, T. Selvam, K. Krithivasan, V.S. Sriram, An efficient intrusion detection technique based on support vector machine and improved binary gravitational search algorithm, Artif. Intell. Rev. (2019) 1–32.

About the authors

Preetha Evangeline David is currently working as an Associate Professor and Head of the Department in the Department of Artificial Intelligence and Machine Learning at Chennai Institute of Technology, Chennai, India. She holds a PhD from Anna University, Chennai in the area of Cloud Computing. She has published many research papers and Patents focusing on Artificial Intelligence, Digital Twin Technology, High Performance Computing, Computational Intelligence and Data Structures. She is currently working on Multi-disciplinary areas in collaboration with other technologies to solve socially relevant challenges and provide solutions to human problems.

Atma Sahu has served for over 30 years at Coppin State University (CSU) in a wide range of positions, including mathematics faculty, applied math researcher, team leader, mathematics and computer science department chairperson, and liaison to K-12 mathematics education. Currently serving as Professor and Chair of the Mathematics and Computer Science department at Coppin State University, Baltimore MD. He has been influential in working collaboratively with colleagues in activities related to Science, Technologies, Engineering, and Mathematics (STEM) research, grant writing, and serving on grant proposal review panels of federal agencies in the United States. He has published numerous research papers in US domestic and international refereed journals in his areas of expertise. He successfully wrote NSF-DST supported workshop grant to increase multi-faculty research capacity involving international collaboration on Elastic Vibrations, Smart Structures, and also, Sahu is an Associate Editor of the International Journal of Fuzzy Computation and Modelling, and an Associate Editor of the International Journal of Engineering, Applied Sciences, and Management. He has earned the 2022 Excellence in teaching faculty award and in 2021 Excellence in Mathematics Research faculty award, from Coppin State University. He served as a university nominee on MHEC advisory council. Currently, he is one of the board of directors of the Indian Institute of Technology Roorkee Foundation, Inc., based in VA, USA.

CHAPTER NINE

Reshaping agriculture using intelligent edge computing

Preetha Evangeline David[a], Pethuru Raj Chelliah[b], and P. Anandhakumar[c]

[a]Department of Artificial Intelligence and Machine Learning, Chennai Institute of Technology, Chennai, Tamil Nadu, India

[b]Reliance Jio Platforms Ltd., Bangalore, India

[c]Department of Information Technology, Madras Institute of Technology, Anna University, Chennai, Tamil Nadu, India

Contents

Abstract

Edge computing in the era of digital transformation is slowly gaining momentum across many industries. It is expected to reach around 75% by 2025. Edge computing is adopted by many industries including the agricultural industry. This technology is helping to build the future of agriculture with smart farming. Though cloud

Advances in Computers, Volume 132
ISSN 0065-2458
https://doi.org/10.1016/bs.adcom.2023.08.007

infrastructure has already played an important role in developing the agricultural sector, edge computing wins the race in terms of speed and efficiency. The opportunity lies within precision agriculture when edge computing is applied to smart farming technologies. While using edge computing technology farmers depend on data to obtain improved control over the industry and optimize the efficiency of their operations which results in reduced operational expenses. Agriculture is one of the world's critical industries and has traditionally been slower to advance and adopt modern technologies than other industries. But now changes are taking place as digitization is becoming more attainable. The agricultural sector is realizing the benefits of advanced technologies like AI, process automation, edge computing, IoT, etc. Edge computing is one of the evolving technologies that have the potential to bring transformation in the agricultural sector. Digitization can help in overcoming some of the biggest challenges in agriculture by using sensors, real-time data-driven insights, and actuators. There are numerous use case examples for smart farming and agriculture starting from keeping track of climate changes and regulating the crop or cattle conditions to greenhouse automation and even farm management solutions.

1. Introduction

The agricultural sector in developing countries has multiple challenges, i.e., if we look at the production side, the first challenge which the farmer faces is the productivity challenge, as it becomes evident that developing countries are very low in productivity, i.e., for any crop, the particular developing country could be the highest producer but at the same time its productivity could be very low. Therefore, crop productivity is a significant factor which has to be looked at. The second challenge which the farmer faces is with respect to climate change, industrial pollution, and pest attacks, as they can damage crops substantially. It is deemed necessary for the farmer to mitigate those challenges by adapting to the latest technologies and insurance schemes. The third challenge is related to market connectivity, whereby the farmer produces crops and thus needs to be connected to distant markets as per the crop production analysis and its subsequent data insights. It is essential for the farmers to have all the information on a digital platform and there should be seamless trade between different markets and different places. However, this connection between farmers and the distant market is nowhere to be seen in developing countries. On the other hand, the industry is looking up for exports, which in turn need to be streamlined substantially. From the economic perspective of the developing country, it is a boon if more facility in nature and regulatory support has to be given for people who are looking to export outside their country or who are looking to have a value-added product across the globe. Now, this mismanagement in the agricultural sector can potentially lead to food security risk.

Precision Agriculture (PA) is intended to help and maximize the development of the farming sector and will also help to ensure food security [1]. It is to be highlighted that PA is a high-tech farming technology that observes, measures, and analyzes farming fields and crops. With the advent of PA, on-field sensors can provide detailed levels of data for problems of soil and weather conditions pertaining to heavy metal toxics and climate change. Big data obtained from sensor networks and farm inputs tracking have a significant role to play to increase farm productivity, reduce environmental impacts, and improve human welfare [2]. By combining artificial intelligence-based big data analytics with sensor and image data, an integrated system could be developed for the agricultural domain. Implementing intensive, high-value, personalized management of crops would increase both production and economic performance. The aim of this paper is to highlight the importance of smart sensors and high-performance computing in protecting stakeholders in the agrifood value chain and providing them with unlimited access to a large dataset of various categories in order to track their farms. The challenges and consideration for the farming sector in developing countries are also highlighted.

Smart agriculture is always connected with high volumes of heterogeneous data sources such as autonomous tractors, harvesters, robots and drones, sensors, and actuators. Heterogeneous sensors and other devices collect relevant agricultural data such as humidity, temperature, pH, and soil conditions. Similarly, it considers the use of various actuators, such as water sprinklers, ventilation devices, lighting, automated windows (in glasshouses), and soil and water nutrition pumps that react according to the data. The number of cloud-based agricultural standalone systems and physical systems is increasing on an almost daily basis, helping to achieve a range of monitoring and analyzing objectives.

Moreover, the last few years of publications have shown that modern computing paradigms such as Cloud, Fog, and Edge play a vital role in agriculture. The main applications of Cloud, Fog, and Edge Computing in agriculture are crop farming, livestock, and greenhouses, which are grouped into different application domains. Some of these applications are implemented with the help of IoT-based sensors and devices by using wireless sensor networks (WSNs), and some other applications are developed with combinations of new computing. For instance, Cloud and Fog, Cloud and Edge, Fog and Edge, or Cloud–Fog–Edge and IoT.

2. Smart agriculture

Smart Agriculture or Smart Farming is an emerging concept that uses modern technologies in agriculture and livestock production to increase production, quantity, and quality, by making maximum use of resources and minimizing the environmental impact. This is demonstrated when farmers and all stakeholders related to agriculture use modern technologies and smart devices to monitor their farms, equipment, crops, and livestock. Using these devices, they can also obtain statistics on their livestock feeding and production of crops [3–5].

In recent years, smart farming has become helpful to all agricultural stakeholders from small to large scale. Smart farming provides benefits not only to scientists and agronomists but also to farmers to access modern technologies and devices that help in the maximization of product quality and quantity while reducing the cost of farming [3]. Smart farming mainly focuses on soil fertility, energy, grassland, water, feed, inputs and waste, machinery, and time management [6,7].

The integration of modern technologies with agriculture achieves the objectives of smart agriculture such as efficiency, sustainability, and availability [8], increased production, water-saving, better quality, reduced costs, pest detection, and animal health [9,10]. The other aims are to increase the reliability of spatially explicit data [3], make agriculture more profitable for the farmer [3], and offer the farmer the option of actively intervening in processes or controlling them [11]. Moreover, big data analysis is another goal of smart farming. Big data consist of massive volumes and a wide variety of data that are generated and captured by agricultural sensors and actuators. In particular, data collected from the field, farm, animals, and greenhouses include information on planting, spraying, materials, yields, in-season imagery, soil types, and weather. Big data analysis provides efficient techniques to do a quality analysis for decision-making [12]. In the coming years, smart agriculture is projected to create a significant impact on the world agricultural economy by applying all modern technologies.

Edge AI will transform the agriculture industry. In most cases, farmlands are localized where there is no availability of high-speed bandwidth, inappropriate resources to handle data, and it is also found that the farmers are not adequately educated about the best practices of agriculture.

AI practices as follows:

1. Soil quality: Examining the soil moisture using a mobile device by checking the farm location and the soil color.
2. Milch animals' health: Tracking the health of livestock by tagging the sensors will give the temperature, heart rate etc. and provide insights about the health condition.
3. Crop Health Analysis: A predictive computation engine, such as drones, can be used to check the health of leaves based on color and the pores it has, whether attacked by insects, pests, or rodents.
4. Disaster protection: Using edge computing, agriculture IoT systems can make informed decisions about potential environmental hazards or natural disasters.
5. Examining leaves health: A predictive computation engine, such as drones, can be used to check the health of leaves based on color and the pores it has, whether attacked by insects, pests, or rodents.
6. Analyzing satellite imagery: Deep analysis of satellite images provides an understanding of agricultural systems. With the help of Geo-spatial data, farmers can get information on crop distribution patterns across the globe along with the impact of weather changes on agriculture, among other applications.
7. Assess crop and soil heat: Predict the effect of different microbes on the health of plants and identify genetic changes that may cause due to harmful pathogens for the plant, among other things.
8. Predictive Analytics: Predictive models in AI help to do seasonal analysis, represent different market scenarios and optimize business costs.

So, here are some of the opportunities that can be brought in by edge computing:

2.1 Ag robots

Autonomous tractors and robotic machinery can run automatically without the intervention of humans, and this can be done with the help of edge computing. The tractors can communicate with nearby sensors to acquire necessary data about the surrounding environment. Ag robots can evaluate the most efficient ways to cover the required area taking into consideration the type of task performed, number of vehicles existing in the field, size of apparatus, etc. Edge computing will enable the ag robots to use computer vision and pre-loaded field data and to interpret the insights from that data. Additionally, the automated tractors can reroute automatically if there is any obstacle like if there is any animal or human in the way. Such smart implements can execute a broad range of tasks, like watering, weeding specific field areas when needed, or even autonomously harvesting crops.

2.2 Farm automation

Similar to ag robots, a greenhouse or even a whole farm can be put on autopilot using IoT edge computing. This indicates that the whole ecosystem can perform the tasks itself without depending on a remote server to process the accumulated data and make decisions about day-to-day processes like feeding the cattle, watering the plants, controlling the temperature, humidity, light, etc. Edge computing will enable the farm or greenhouse to work without depending on the connection to the main server and make decisions based on data from local sensors. This can result in improved process reliability and reduced waste, making agriculture a more sustainable process.

2.3 Disaster protection

Agriculture IoT systems can make sophisticated decisions about possible environmental hazards or natural disasters with the help of edge computing. Remote sensors can accumulate and examine data regarding the changes in the weather or the environment to forecast possible disasters. If there are definite indications of danger, they can instantly process the information to the general control center. Farmers thus will be able to take real-time appropriate measures to shield their crops. It is expected that the dependence on edge computing by enterprise-owned IoT devices will reach 6.5 billion in the coming year. Agriculture now has all the opportunities to lead the innovation in this field including manufacturing, transportation, energy, retail, and healthcare. This indicates that we can predict more edge computing use cases in agriculture soon.

3. Smart farming initiatives is the need of the hour

With the advent of Internet of Things (IoT), smart devices have reached into all facets of our day-to-day life, i.e., healthcare and wellness, smart homes, automobile and logistics, intelligent cities and industries. In recent decades, agriculture has seen a series of technological changes, increasingly industrialized and technologically driven. Through different agriculture-based smart devices, farmers today now have greater control over animal husbandry and cultivation processes, making them more predictable and productive. This, along with the rising market demand for agricultural products, has helped to increase the worldwide proliferation of intelligent agriculture technologies.

Modern agriculture can be addressed in several respects. For instance, AgriTech refers to the use of technology in the domain of agriculture. In addition, intelligent agriculture is primarily used to describe the use of IoT-based agricultural solutions. With IoT sensors, farmers can make informed decisions and develop various parameters of their work, i.e., cattle to crop production, in order to collect environmental and machine metrics. For example, farmers can decide exactly how much pesticide and fertilizer is to be utilized to optimize productivity by using smart agriculture sensors for monitoring crop status. The same applies to the concept of intelligent farming. Fig. 1 shows a broader perspective on a modern-day agricultural model, which incorporates various wireless sensor nodes to enable IoT-based farming with satellite communication, where different ground sensors are deployed which communicate with the cloud computing node for data processing and analysis, so that farmers can make correct decisions.

Although smart IoT and industrial IoT are not as common as consumer-connected devices, the market continues to be very competitive. IoT technologies are increasingly being implemented for agriculture. COVID-19 has had a positive impact on IoT market share in agriculture. Indeed, the smart framing market share is expected to hit $6.2 billion by the end of 2021, as reported recently.

It is evident that COVID-19 has made a significant impact on the farming sector across the world. However, the agricultural sector is showing potential to make a strong comeback by leveraging positive government policies which indicate adoption of advanced technologies by making substantial investment in the agricultural sector. This initiative will make room for IoT-based agricultural solutions as a prominent business strategy, thus causing a reasonable increment in crop production. However, in the current situation, the market is expected to show a decline up to 0.8% for the first two quarters of year 2021 compared to 2020 and this trend will show a positive growth from 2022 onwards.

In addition to this, the smart world agriculture market is projected to triple to 15.3 billion dollars by 2025, compared to just over 5 billion dollars back in 2016. If the sector continues to expand, there will still be plenty of opportunities for companies. In the coming years, creating IoT products

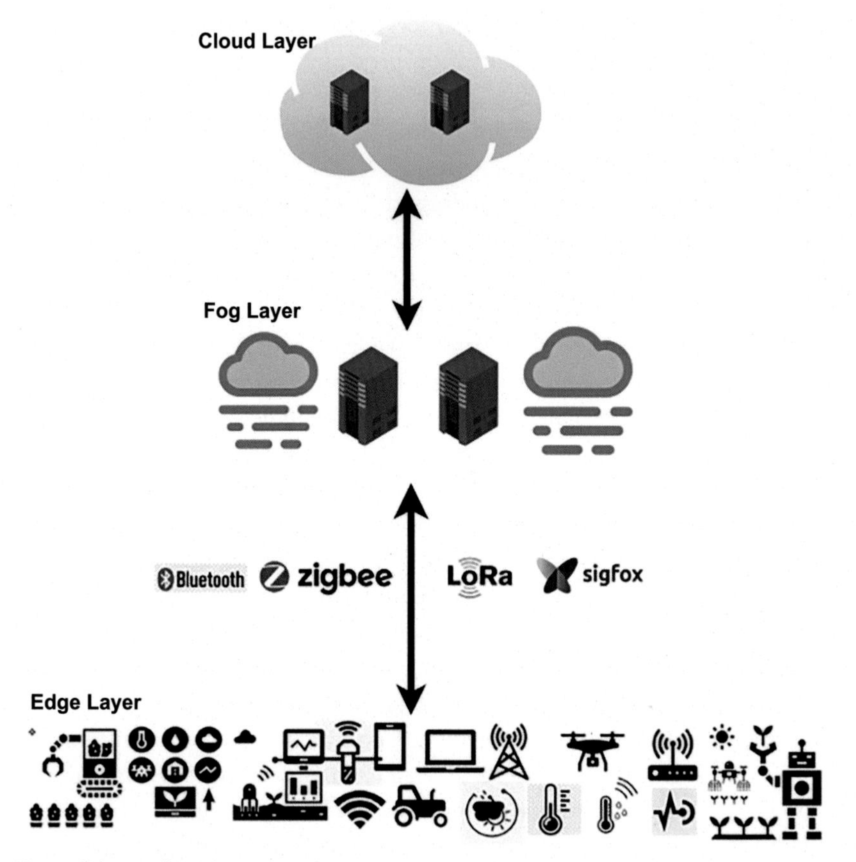

Fig. 1 Edge architecture.

for agriculture will distinguish companies as early adopters, thus helping to pave the way for success.

This model can be used for any other application domain with some minor changes based on the domain requirements. In the proposed architecture, the Cloud layer is mainly for ample scale data storage and data analytics. This layer is also responsible for loading algorithms and data analytical tools to Fog nodes. This can also be used to store backup data for future analysis. The Fog layer is essential in this model, and this will be installed in local farms. Fog layers will be responsible for real-time data analytics such as predicting pests and diseases, yield prediction, weather prediction, and agricultural monitoring automation.

Moreover, this will make decisions on real-time data and do reasoning analysis as well. Finally, the processed and analyzed data can be uploaded to the Cloud layer for backup purposes or further analysis. The third layer is the Edge, consisting of end devices, tractors, sensors, and actuators. The main goal of this layer is the collection of data and its transfer to the Fog layer.

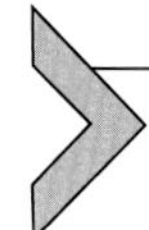

4. Commonly used sensors for smart farming and heavy metal identification

Sensors for Soil Moisture (SM) have been used in crop fields for decades to measure water content. The use of handheld/manual soil moisture technology is increasingly being replaced by automated technologies, since there were difficulties in manual soil moisture readings in remote production areas. In the past decade, technology has been developed for wireless data collection, providing managers and users with real-time access to soil moisture data, resulting in more successful water management decisions. Some of the prominent sensing devices to measure soil moisture comprise gravimetric sampling, resistive sensors, capacitive sensors, and Ground Penetrating Radar (GPR). Gravimetric sampling is a direct and normal SM measurement tool. SM is determined by a proportion of dry soil mass to wet soil mass including pores. It needs the manual drying of soil samples taken from the field and oven sampling. The electrical conductivity of water and the measuring of resistance changes based on soil water content are primarily resistive sensors, such as granular matrix sensors. This method includes sensor calibration for precise SM reading.

Intelligent irrigation-based measurement to maintain soil moisture levels is significant to improve plant productivity and quality. On the other hand, soil moisture sensors these days are expensive, i.e., the ECHO-EC5 soil moisture sensor costs around USD 169. In order to overcome the cost constraint factor, Wang et al.proposed an RFID-based GreenTag sensor to maintain and improvise plant productivity and quality.

In addition, RFID sensors can be combined with biosensors comprising aptamer and DNA-based properties which can be used to detect heavy metals at nanoscale and large scale levels pertaining to food safety monitoring. A heavy metal detection-based biosensor is composed of genetically modified bacterial cells and a green fluorescent signal amplifier which detects the presence of arsenite in foods. Its arsenic detection lasts for an hour with a detection range of 5–140 μg/L. Other methodologies pertaining to biosensors, i.e., aptamers and graphene electrodes, have also been used to detect arsenic with the possibility of being developed as simple and easy-to-use low-cost devices.

The EC-5 series sensors were also used by Wu et al. for field-specific calibration and evaluation in sandy soils. Nonetheless, EC-5 sensors have turned out to be helpful to reveal soil water content dynamics in different soil depths post rainfall conditions. The ECHO series has other variants of sensors; i.e., ECHO-EA10 can be used for medium textured soil type with low electrical conductance conditions. In addition to this, there is ECHO-10HS soil moisture sensor which is a new addition in the soil moisture sensor family and possesses high-frequency oscillation, which enables the sensors to accurately measure soil moisture in any of the soil or soilless media with minimum salinity and textural effects.

In order to measure soil water content and salinity, Zemni et al. used 5TE sensors at different soil depths to assess dielectric permittivity (Ka) and electrical conductivity (ECa). It is to be noted that 5TE sensors are based on frequency domain reflectometry (FDR); therefore, they use a fixed frequency wave of broadband signal which makes the device cheaper and more compact. Nolz et al. deployed hydro probe2 sensors to evaluate near surface soil water and determine in situ water retention function. Hydro probe sensors are advantageous due to their linear signal response. On the contrary, hydro probe sensors are not suitable for sandy soils. Udukumburage et al. used an MP406 soil moisture sensor to verify the saturated condition of the expansive soil layer. They also used this sensor to measure volumetric water content values in the soil column during the wetting and drying process. In order to maintain the indoor ecosystem services, air quality plays an essential role. In this regard, MIKROE gas sensors are used to monitor the air quality. To evaluate and assess the vegetation change and study physiological and metabolic response of corn fields and paddy fields, the Pogo II VWC has been widely used.

Hu et al. used Portable X-ray Fluorescence Spectroscopy (PXRF) to assess the heavy metal content in soil for which they covered 301 farmland soils from Fuyang in Zhejiang Province, in the southern Yangtze River Delta, China. Conventional methods for heavy metal detection such as Atomic Absorption Spectrometry (AAS), Atomic Fluorescence Spectrometry (AFS), and Inductively Coupled Plasma Optical Emission Spectroscopy (ICP-OES), are expensive and lengthy procedures which are executed in laboratories. Therefore, these methods are not taken into consideration for rapid testing and high-density evaluation of soil heavy metals contamination. As an alternative method for rapid heavy metal detection, Portable X-ray Fluorescence (PXRF) was used to assess cumulative concentrations of soil heavy metals based on linear regression models between fluorescence intensity and specific heavy metal concentration. Due to its ease of use and rapid testing ability using non-destructive quantification, PXRF has been widely used by researchers in numerous domains. For the heavy metal assessment in agricultural soil conducted by Hu et al. VNIR sensor was used to anticipate soil properties comprising pH, soil nitrogen, and carbon. In addition to PXRF, NixPRO color sensor can also be used to identify hotspots and total spatial area in excess of environmental thresholds in landfill soils.

Lately Zhao and Liu have developed a Portable Electrochemical System (PES) for on-site heavy metal detection on farmland. Their system was composed of a three-electrode configuration which comprised a signal acquisition system integrated with a microcontroller-based potentiostat to perform square-wave anodic stripping voltammetry. Their system was assessed by testing the detection of pd.(II) and cd(II) in acetic acid soil extracts and acetate buffer solution. However, their system did not include any wireless sensor module to transmit heavy metal composition data.

Other than the aforementioned sensors, there are several other wireless sensors dedicated to: photosynthesis, i.e., Beta Therm temperature sensor; leaf wetness sensor, i.e., SLWA-M003; precision sensor for leaf temperature, i.e., ΔLA-C; light intensity sensor, i.e., BH1750FUI sensor. With the advent of these sensors, CO2 sensors also play an essential role, especially in greenhouse systems. CO_2 sensors have also been widely used to measure the subsequent level in peat soil, landfill, and forest control site. In the smart farming ecosystem, the growth and quality of the fruit bunch cannot be neglected. In this regard, there are dedicated fruit growth monitoring sensors which researchers have used in their domain of plantation. Thalheimer designed an optoelectronics sensor for monitoring fruit and stem radial growth. Their developed sensor was lightweight and easy to install with low maintenance. Nonetheless, the sensor was well tested in open field conditions. In addition to this, the effect of gas concentration during the fruit growth was studied by Ma et al., for which a smart ethylene electrochemical sensor was established to investigate ethylene emission from fruits. Lately, Hanssens et al. came up with a heat field deformation sensor to measure sap flow dynamics through the tomato peduncle. Heat griddling of the peduncle was performed to differentiate flow of xylem and phloem with respect to developing fruits.

Capacitive sensors calculate SM on the basis of changes in soil capacitance due to differences in water content. Commercial UTs use capacitive sensors, which are usually more accurate than resistive sensors but cost more.

Ground Penetrating Radars (GPR) are based upon electromagnetic wave absorption and reflection. SM sensing uses impulses, frequency sweeping, and frequency-modulated technologies. This method is used for measuring soil moisture near the surface (up to 10 cm). The most reliable soil humidity samples used in fields are neutron scattering samples and scattering samples use radiation methods for calculating SM by estimating changes to the neutron flux density due to water content of the soil. However, in such cases, specific licenses are required to carry out its implementation.

Numerous research studies have been performed to develop electrochemical devices for various applications, which are known as potentiostat. Lately, an Arduino-based potentiostat was fabricated from cost-efficient components and was able to execute simple electrochemical experiments, whereby the results were recorded and analyzed in a Windows operating system via USB interface. As an addition to Arduino-based potentiostat, Raspberry Pi (RPi) controller was also used to execute the electrochemical experiments, whereby the results were displayed on the LCD touch panel connected to the controller. Both Arduino- and Raspberry Pi-based potentiostat have the potential to incorporate wireless sensors for data transmission; however, these controllers do not contain a built-in Analog to Digital Converter (ADC) and Digital to Analog Converter (DAC) which make the overall design more sophisticated. In this regard, Hanisah et al. came up with a portable Heavy Metal Potentiostat (HMstat) to detect heavy metal composition on-site. Their potentiostat comprised a digital Control Signal Component (CSC) and the electronic component, which is the analog Potentiostat Read-out Circuit Component (PRCC), Nonetheless, it is worth noting that both the Arduino and RPi controller board do support the incorporation of various sensor modules. Therefore, researchers have room to incorporate soil moisture and temperature sensors along with other sensors depending on the slots available in the controller; thus, an integrated system for soil moisture and heavy metal analysis can be developed.

Other soil physical properties can be calculated to populate the map of the soil with other soil properties such as soil organic content, pH, sand, silt particles percentage, and nutrients such as Mg, P, OM, Ca, base saturation Mg, base saturation K, base saturation Ca, CEC, and K/Mg. In situ, calculating these properties in real time also faces challenges due to scale, cost, and technology limitations.

In precision farming, some of the long-lasting decisions can be taken using yield monitoring. This method helps in providing spatial distribution of crop yields at the end of the growing season. Yield sensors are normally mounted on farm equipment and capture yield data automatically in the course of the harvest. In particular, mass flow sensors on grain containers are mounted to record grain inflows along with the position. The collected data are analyzed with tools such as ArchInfo, Mapinfo, and Environment System Analysis International.

In order to get an insight into the crop yield combined with field topography, Electrical Conductivity (EC) sensors are used. Soil's ability to conduct current is measured by electrical conductivity. EC assessment is used to assess the use of phosphorus, cations in water, drainage, and rooting depths. EC maps are used for zoning the area. The zoning is also used to incorporate precision agricultural practices such as variable rate irrigation, variable rate seeding, and drainage management. Electromagnetic Induction (EMI) methods can be used for the mapping of the EC by apparent Electrical Conductance (ECa) and Visible Near Infrared Reflectance (VNIR). There are a number of commercial tools available, i.e., Veris 3100, EC400 sensors in conjunction with GPS systems.

In the domain of soil sensing, macronutrients such as nitrogen, potassium, and phosphorus are essential to the growth of crops. The evaluation of these nutrients helps to assess the effects of fertilizer and potential applications. The optical detection is based on reflectance spectroscopy to measure the macrosimulation's reflection and absorption. A sensing system using planar electromagnetic sensors has been developed in the detection of nitrate and sulphate concentration in natural water resources. This approach is used to detect the amounts of nitrate and sulphates by correlating the impedance of the sensor array with their concentration. The key approaches to soil macronutrients include electrochemical, VIS-NIRS, and ATR spectroscopy. These approaches to soil macronutrients are limited to sensing a single desired ion because the membrane used in these methods only reacts to one ion. To achieve a simultaneous multi-ion sensing, it is necessary to build a detector array for the sensing of soil macro nutrients.

There are several opportunities to advance the state of precision farming through the utilization of the above discussed sensors.

5. High performance computing on edge (HPCE)

This new High-Performance Computing (HPC) solution seeks to move beyond the agricultural services offered on edge and provide a comprehensive platform for precision farming and animal husbandry and furnish with utility not only for farmers but also for stakeholders. The HPCE architecture is adapted from CYBELE conceptual framework. The HPCE model uses open and proprietary vast amounts of datasets, including sensor readings, as well as satellite data and historic climatic and environmental information for ready reference. While this would be the most effective way to use HPC technology, it only uses the latest software platforms and projects that are being developed by HPCE's e-controlled services, as well as increased HPC e-infrastructure to enable huge heterogeneous data processing to be done and find modern solutions to complex problems using dedicated algorithms. Due to the interconnection of large-oriented approaches, varying datasets, and available big data techniques, it is possible to scale distributed big data research to enormous scales when holding many types of datasets together in one place. In doing so, it enables the aggregated data and metadata to be aligned semantically to a standard scheme and data model and enables advanced data analytics to take secret information into account. In addition to this, the HPCE architecture will also help in gaining insights from adaptive data visualization services.

With reference to CYBELE, the architectural approach of the HPC on edge and by organizing a product component based on interdependencies, this is intended to highlight the importance of pipelines being constructed to promote compatibility and show how to maintain the integrity of interdependent services. It is worth noting that CYBELE resonates well with the EdgeX platform architecture. EdgeX platform comprises four core services, i.e., device services, core services, supporting services, and application services to enable smooth workflow optimization. In addition, it will be interesting to see a synchronization of EdgeX with a dedicated HPC framework for faster batch processing of data over edge.

Big, heterogeneous data are made available through repositories powered by HPC which is responsible for the processing at the edge layer. In this regard, HPC frameworks such as Spark, Hadoop, YARN, Big Deep Learning (BigDL), Directed Acyclic Graph (DAG), and Kubernetes are deployed for the batch processing of data using distributed framework attached to the edge layer. It is worth noting that Spark and BigDL are the widely used frameworks in many organizations for their open source and high degree of interoperability features. Spark and BigDL are based on MapReduce framework which has high room for tuning for smooth workflow optimization. The transmission of the application process interface along with data from the cloud layer to the edge layer is conducted using 4G/5G or fiber/DOCSIS/DSL communication system. This is seen on the middle section of the architecture. At first, the data are processed in the background prior to being passed on to the check-in stage for data validator or timestamp validator for resolution of data verification and timing problems. Once data are obtained in edge layer, quality checks are conducted to identify anomalies and any other data irregularities, maintaining their accuracy and validity, which are accompanied by a series of measures aligned with processes of data cleansing.

Finally, the HPE data provenance service provides the mechanisms required for recording all relevant information concerning incoming data of interest. With HPE, the data provenance platform is inherently connected to the data policy and asset brokerage engine that enables the platform to bind data providers and data users with data share and business features. In addition to facilitating interoperability and reuse of data, the inspected data are annotated and harmonized semantically. Since the data come from a variety of physically distributed data sources, a standard data model will be created for the semantic definition and annotation of the data. To facilitate the pipeline and allow the various heterogeneous components to communicate seamlessly, the model will be used as a common language to annotate data and exchange messages between the components. Clean and semantically uplifted data are then available, i.e., open and proprietary data to be queried, analyzed, and viewed. An exemplification of how ground sensors have their data stored and analyzed at cloud data base. The on-field data are continuously assessed by a real-time monitoring system to ensure triggering effects if any threshold point is crossed. Simultaneously, the on-field data are also stored in the cloud database from where the user can download the required data and at the same time, data analysis could be applied using the machine learning tools stored over the cloud database.

To facilitate simulation execution, a defined experimental composition setting is designed, as shown in the top right part of the architecture (cloud layer). The composition framework of experiments aims to support the separate design, development, and execution of big data research procedures, the support of embedded scientific computation and reproductive tests. In the analysis method, its subsequent template is selected to provide each analytical template with its own software and execution endpoint and allow the user to modify the appropriate configuration variables (i.e., input algorithm, execution parameters, netting parameters, and output parameters). The results of each analytical template are presented. The composition system for experiments will promote the design and implementation of data analysis workflows consisting of a number of data analysis procedures, interconnected in terms of data sources and input and output artifacts.

The outcome will constitute the input to another research template when a template is executed. The output of the research model is an object for session that contains all the memory output values.

In addition to big data, advanced analysis must be implemented when selecting input datasets and developing workflows. For HPCE, advanced analytical algorithms are available to stakeholders that allow them to explore various forms of data visually and to find and solve new trends. In order to achieve improved delivery and monitoring, machining and predictive modeling methods should be modified so as to handle the predictive life cycle of data planning, detection, and analysis. However, the implementation of advanced analytics along with huge, complex data increases the need for strong computing power and a higher processing memory, so that information can be collected within a realistic timeframe. When the test cases are executed, multiple HPC attributes are needed, including storage power, speed of the device, memory capacity, and quick turnaround time.

The next section discusses the IoT-based communication methodologies in edge computing used for precision farming developed by several researchers.

6. Processing in agriculture

Edge Computing is like a specialization of the Internet of Things. Without it, all data collected through IoT devices are sent to a cloud centre for processing. With the new technology, on the other hand, the collected data is classified locally, so that part of it is processed right there, on the "edge" of the network, hence the name edge computing through micro data centres.

Thus, only certain information is sent to a cloud centre, while those that often need to be consulted are analyzed on the device itself in the case of agribusiness, in the field reducing data traffic.

This ability to perform advanced analytics close to the data source meets the market's need to cope with increasing traffic demands. As there is a screening of the information that will be sent to the processing centre, transfer rates are optimized.

The main benefits of Edge Computing are the reduction in bandwidth required for sending and processing data and the decrease in latency, which is the response time of a request — the period in milliseconds that a data takes to navigate from where it was generated to where it will be processed. Both advantages are possible due to the proximity between the processing location and the origin of the information; only with cloud computing, on the other hand, all data would need to travel long distances before returning to consumption.

In practice, this reduction in latency helps in real-time data access, which is essential for the implementation, with maximum effectiveness, of digital and intelligent solutions in agricultural processes. Certain functions can be performed on the equipment itself through Edge Computing, making it easier to make smarter and more agile decisions.

Another example is the use of this technology in an agricultural spraying activity, in which sensory devices are enabled to determine alone which area should be sprayed, using the data collected and analyzed by the devices themselves.

As Edge Computing also reduces the bandwidth required for processing, the solution becomes even more useful for agricultural applications. In the current context, in which IoT solutions already allow wide integration between various products, such as sensors, on-board computers, edge computing machines, is playing an increasingly important role in the application and evolution of technology in the field. It is one of the technologies that will have increasing adoption in the coming years, accelerating the consolidation of the digital transformation of agribusiness.

7. The proposed system

The proposed learning model for irrigation is implemented in a prototype IoT system that has four components: (i) Edge node layer — This layer consists of sensors, actuator, and two microcontrollers. In this layer, edge node acquires the sensor data from the surroundings and controls the actuator for actuating water pumps to start irrigation. (ii) Edge server layer — This layer consists of Raspberry Pi that act as edge server and capable of multitask processing. Here, edge server controls the edge nodes for sending signal and receiving data at regular interval of time. It is also connected to the cloud server for receiving developed and trained machine learning model to be deployed and make irrigation decision for controlling edge nodes. (iii) Edge service layer — This layer is deployed in the edge server and it is responsible for controlling the whole system through a developed web dashboard. The dashboard has live feed data, control of edge nodes, and cloud services access. This service layer also has the control access of the proposed machine learning model. (iv) Cloud server layer — This layer composed of cloud services and cloud storage where its role is to train the

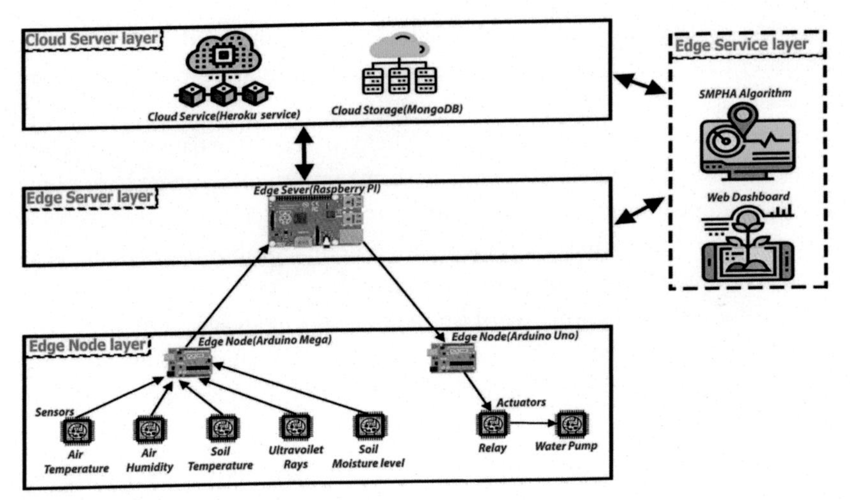

Fig. 2 Proposed smart irrigation system.

machine learning model and store the data in database. It sends the trained proposed model to the edge server for decision-making regarding irrigation scheduling. The comprehensive interconnections in the system are shown in Fig. 2.

The proposed IoT-based smart irrigation system includes five major components: field deployed module, Web-based interface, Web API weather input, soil moisture prediction mechanism, and edge communication model.

7.1 Web-based interface

The proposed framework consists of a web-based application to allow farmers visualize the growing data and interacting with the garden in real time. In addition, users can also be able to examine and analyze the historical growing data, if needed, through functionalities such as irrigation control, motor control prediction model deployment, and manual data entry implemented in this web application. Here, Node.js was chosen for developing the web application [13,14], while Huang [15] was utilized as the database system. Data stored in the database, which is deployed in the cloud, will be used for further data analysis in the future. The web application's functions are designed as a software design pattern called model-view-controller (MVC).

In the frontend, ChartJS is used to represent data through dynamic charts. The web application is also used as an interface to manage all the physical devices/actuators in the garden. To deploy the web-server to the cloud, a cloud platform as a service (PaaS), namely, Heroku, had been utilized. Heroku is a cloud platform that provides platform as a service (PaaS), facilitates the creation of applications and deploying these online rapidly [16,17]. It also enhances scalability and functionality by integrating several add-on services. The field data are sent to the server by Raspberry Pi using this web service. This web service manages the network outage/fluctuation during data synchronization from the field device to the server by taking the help of flag settings at the database level. The interface facilitates the scheduling of irrigation along with visualizing real time sensors and predicted soil moisture for upcoming days and precipitation information. By using the denoted threshold value of soil moisture suggested by agronomists, the irrigation can be scheduled by the user.

The system maintains the threshold value depending on the predicted pattern of soil moisture and precipitation information. The process of irrigation is initiated automatically and stopped after the specified threshold value generated from the proposed algorithm of soil moisture when it is reached.

7.2 Edge communication model

The communication protocols in the proposed framework are flexible and transparent in nature for accepting both wired and wireless methodologies. For the maximum utilization of potentiality in edge computing components, the communication among various components in the edge-IoT system requires intense probing by using the versatility among the devices in network edges. For transferring the data gathered from pivot sensors, a communication technology such as Zigbee [18] is needed for the irrigation systems. Therefore, the communication component in the proposed work is classified into three main areas as shown in Fig. 3.

The Message Query Telemetry Transport (MQTT) protocol is used for the communication in the proposed system. The analysis in ref. [19] presented seven IoT messaging protocols (MQTT, CoAP, XMPP, AMQP, DDS, REST-HTTP, and WebSocket) as communication protocols that play a major role in smart farming. The authors have concluded that MQTT proved to be the most secure protocol after probing all the protocols with respect to latency, energy and bandwidth requirements, throughput, reliability, and security. Moreover, MQTT is secure in both end-to-end architecture and gateway server architecture. In an MQTT setup, a

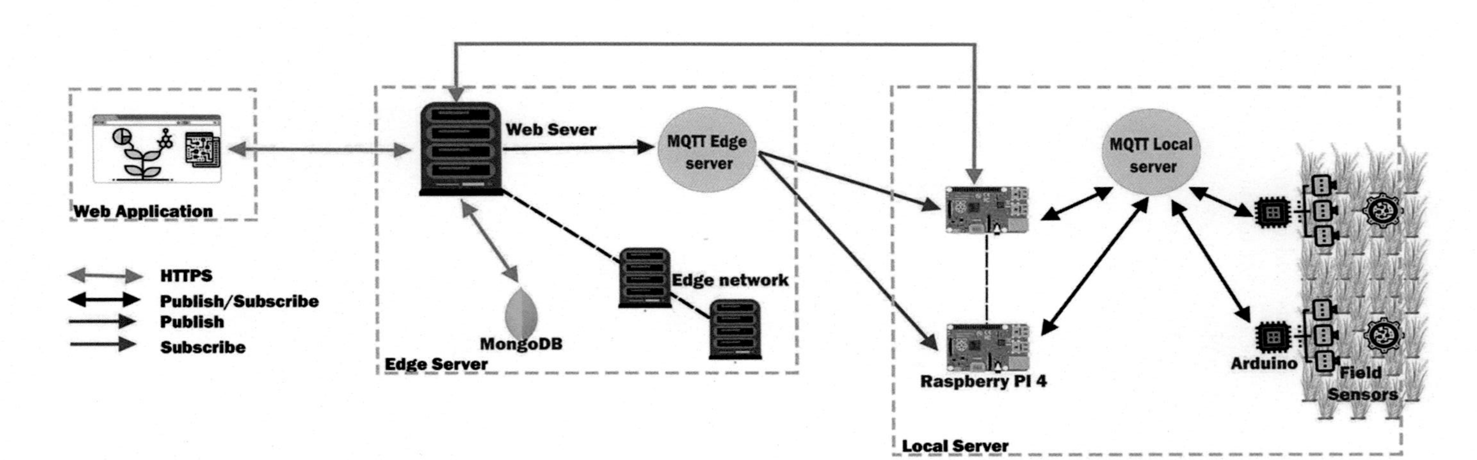

Fig. 3 Edge communication model.

MQTT server termed as MQTT broker executes on the IoT solution [20]. Under a common identifier, a "publisher" and a "subscriber" link among themselves to this broker. In the IoT solution, publishers and subscribers are the IoT devices and IoT hubs or control devices, respectively. When the publishers have new data for recording, the data are published to the broker. The broker then flags that it has new publisher data, and the corresponding data are read by the subscriber. Then, the subscriber analyzes the data and reacts accordingly.

The first level accomplishes with connecting the end users to system with the help of mobile or web-based applications through the Internet. The next level (cloud computing server) deals with the connection of web server and MQTT broker for directing the user requests and other components at the edge landscape or from the farms to the right cloud-based services like displaying the real time status of the farm for the users, triggering a new deployment of the updated ML model to the corresponding edge node. The third level (farming area) is directed toward the deployment of sensors and IoT devices (actuators) for communicating with other components in the entire system.

7.3 Deployment of soil moisture prediction hybrid algorithm

The watering mechanism of the plant has different approaches in the proposed model. Primarily, the system is trained with manual irrigations datasets during the process of learning with respect to suggestions defined by agronomists. The model is trained to learn the needs of irrigation in the first level of deployment in cloud without the inclusion of pre-processed data. After acquiring the required data and training, the proposed system is initiated to grasp the plant's watering needs by undergoing plenty of manual irrigations. Thereafter, manual irrigation is not required and the system makes automated decisions in watering using the gathered data and the application of ML methods. The proposed model then decides the irrigation strategies automatically using ML methods without the need including collected datasets in the automatic irrigation process. The proposed model can be improved through the learning process when the number of precise irrigation inputs is provided to the model at each stage of training.

The decision-making procedure is developed with two modules for irrigation strategies according to the soil moisture prediction for upcoming days. The first module deals with training the model in cloud with manual irrigation datasets through steps such as data collection, data preprocessing, training, and model development. The system acquires values of air temperature (TH), soil temperature (SMT), soil moisture (SM), humidity (HU), and ultraviolet rays (UV) periodically from the physical environment in the data collection stage, which is essentially required for arriving at the watering decisions. Also, the time of performing the manual irrigation is recorded in the database. These data are timestamped and stored in as datasets to aid in making decisions for knowing the time of irrigation. In the next step of pre-processing, inconsistencies are eliminated and outliers caused by sensor errors are detected from the irrigation dataset, thereby helping in the removal of broken data. The training stage involves the application of supervised machine learning (ML) algorithms. Here the regression algorithms such as support vector regression (SVR), multiple linear regression (MLR), lasso regression (LR), decision tree regressor (DTR), random forest regressor (RF), and XG-boost regressor (XB) techniques are used for the deployment. The regression algorithms are trained using the collected datasets. Finally, through training, regression models are created, namely, SVR model, MLR model, LR model, DTR model, RF model, and XB model that are been combined with the second module for decision-making.

The second module caters to the prediction of irrigation for upcoming days by infusing the weather data as an input to the regression trained models. The live datasets from the weather API for future prediction of soil moisture variable are used. The dependent variables from weather forecast data like temperature (TH), humidity (HU), ultraviolet (UV), and precipitation (PC) are tested in the aforementioned model for soil moisture prediction. Then, the regression trained model is evaluated and deployed using the weather testing data for the prediction of soil moisture in accordance with the precipitation. After the prediction of data for the upcoming days, these developed regression models are combined with unsupervised ML algorithm named k-means clustering for estimating the changes incurred in soil moisture prediction due to the impact of weather conditions. Further, each regression models with k-means algorithm are evaluated for performances in terms of irrigation decision-making process as shown in Table 1. The combined algorithms are estimated through MAPE, MSE, R2, execution speed, power consumption, and accuracy. The estimation and computation of these parameters are detailed by the authors in ref. [21].

Table 1 Comparison of performance metrics obtained from various ML algorithms.

Algorithms used	Accuracy	R^2	MSE	MAPE (%)	Execution time	Power (J)
SVR + k-means	0.96	0.96	0.25	1.98	0.06078	1164.85
MLR + k-means	0.94	0.88	0.31	2.15	0.02075	429.30
LR + k-means	0.95	0.94	0.32	2.23	0.02482	351.35
DTR + k-means	0.93	0.95	0.29	1.62	0.15687	914.70
RF + k-means	0.95	0.91	0.27	1.57	0.16745	1475.13
XB + k-means	0.97	0.98	0.20	1.08	0.03547	537.87

Table 2 Comparison of predicted SM value with actual SM value.

Date	Average SM value from sensor	Average predicted SM value (XB + k-means)
28-09-2022	35.23	34.04
29-09-2022	36.41	37.20
30-09-2022	31.57	30.46
01-10-2022	34.66	33.15
02-10-2022	36.73	37.12
03-10-2022	32.88	33.01

XGBoost + k-means (XB + k-means) approach provides more accuracy with less MSE comparatively and also the R2 with 98% in soil moisture prediction using combined approach is given in Table 1. It is evident that the proposed combination performs better when compared to other regression + k-means-based approaches. XB + k-means-based hybrid machine learning algorithm is applied in irrigation planning module on account of aforementioned performance metrices of ML. Although it performs moderately in terms of execution time and power usage, it is selected for the deployment in edge computing as it has better performed in terms of accuracy, R2, MSE, and MAPE metrices. It is observed that the prediction of soil moisture for the upcoming days from the proposed algorithm (XB + k-means) is nearer to the actual value as shown in Table 2, and hence, XB + k-means is selected for the implementation of SMPHA in edge-based irrigation scheduling.

7.4 Edge layer setup

The edge node acts as a computing center where incoming data are analyzed and fed as the input vector to the ML model for processing and to return the control signals for activating or deactivating the actuators placed at the farm. Edge node processes the physical data (real time) at every end device such as the collected and processed data via the Raspberry Pi nodes presented in the proposed scheme. The prediction model is designed using TensorFlow API and trained, tested on Google Colab in this work. Amazon Web Service (AWS) offers a library named Boto3 having many APIs to upload and download objects. After the development of model, it is transferred to Amazon S3, a service provided by AWS. The edge node utilizes the trained model from S3 for analyzing the sensed data acquired from garden's sensors. The decision is delivered based on real time data analysis at the edge node and transmitted to Arduino nodes in the fields landscape immediately for controlling the actuators. In another flow, the data collected from sensors are filtered so as to keep only the modified data at the edge node before being sent back for mitigating the communication cost to the database in the cloud. These data are used in the updation of the ML model to enhance its efficiency.

7.5 Analytics setup

The main goal of this experiment lies in gathering the various physical parameters of a farming land via sensors and utilizing the fetched data along with weather forecast information for developing an algorithm using hybrid machine learning approach to infuse higher accuracy in predicting the soil moisture for the upcoming days. As discussed in Section 4, for the proper planning and provisioning of optimal irrigation, the algorithm provides a predictable estimate of soil moisture with the assistance of various statistical measures as shown in Table 1. The measures are adopted for estimating the appropriateness and error rate of the proposed algorithm. It is inferred from the experiment that, optimal irrigation is feasible using a good estimation (close to the actual value) of the soil moisture (Table 2), with the support of field data and forecast information, thereby utilizing the natural rain efficiently.

7.6 Work flow

The flowchart in Fig. 4 depicts the working of the proposed system based on the decision support system that is beneficial for irrigation needed for the growth of vegetables. The chili plant is grown in a growbag attached with

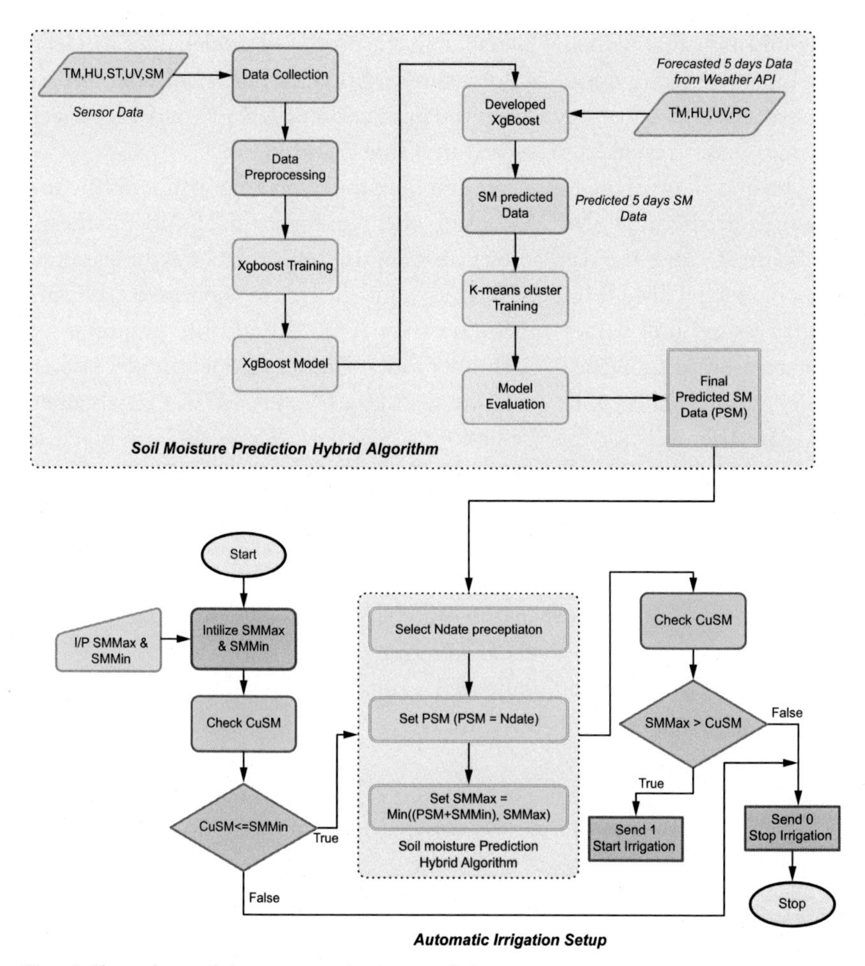

Fig. 4 Flow chart of the proposed edge model.

sensors and Pi and monitored for 95 days of data collection. To bring out optimality in the irrigation system, features relating to climate, soil, crop, and field infrastructure are to be considered. To provide several recommendations in the production of vegetables, decision support systems (DSSs) are designed, which process voluminous information [22]. This proposed work is the extension of soil moisture differences (SMD) model [23] developed for soil moisture prediction. The threshold values of soil moisture are used in the SMD model where the system schedules the irrigation date based on the predicted soil moisture and weather forecast (precipitation) information automatically using SVR+ k-means modeling. Therefore, in the extension of the aforementioned work, further more number of sensors are used to log soil moisture value, which is averaged in the proposed model. This model is developed in two divisions of flowchart as shown in Fig. 7, where both are interconnected. It is observed that the prediction of XB + k-mean approach provides better results as presented in Table 2 and Fig. 5.

The first phase of the flowchart describes the hybrid algorithm for the soil moisture prediction (SMPHA) using the combination of XB + k-means algorithm. During the data collection step, the sensor data for the parameters, namely, TM, HU, ST, UV, and SM, are collected. During preprocessing, null values and outliers are removed and the preprocessed data are used to train the XG-Boost model. The developed model is then trained with variables of live weather features (TM, HU, UV, PC) obtained from Weather API for the prediction of SM data. These data are given as input to k-means clustering algorithm to predict the soil moisture, which is defined as SMPHA value to be infused in the next phase of the flowchart. The second phase of the flowchart defines the automatic irrigation planning setup. The setup starts obtaining the soil moisture maximum (SMMax) and soil moisture minimum (SMMin) values in the dashboard for setting the maximum and minimum level of soil moisture. Then, the current soil moisture (CuSM) is sensed and compared against the threshold SMMin. If the resulting value is less than SMMin, the process proceeds with SMPHA.

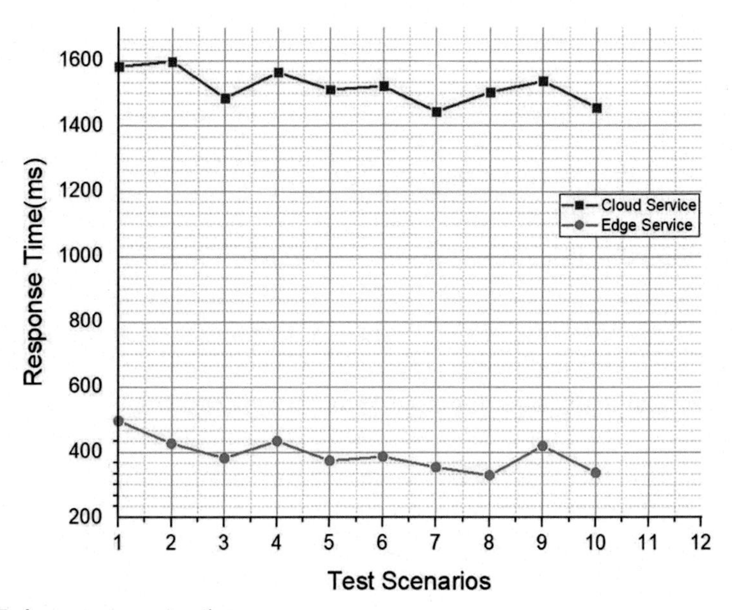

Fig. 5 Average response time.

On the contrary, it stops the irrigation process by sending 0 to the relay. In SMPHA, the nearest precipitation date is selected and it is assigned to the predicted soil moisture (PSM). The SMMax is decided by finding the minimum of (PSM + SMMin, SMMax), and the predicted SMMax is further checked against CuSM with a condition if SMMax is greater the CuSM then it sends 1 to the relay as a signal to start irrigation. If the condition fails, then it sends 0 to stop irrigation. The process of automatic irrigation ends by forecasting the irrigation schedule in accordance with the live weather parameters.

7.6.1 Evaluation

A hybrid machine learning methodology is used in evaluating the first stage of the proposed model. The predicted value of the soil moisture is better in terms of their accuracy and error rate. From the comparison of the other ML algorithms as shown in Table 2, XB + k-means performs better and taken further to be deployed in edge and cloud to check its efficiency with each other. Therefore, for analyzing the efficiency of the edge server in accordance with the proposed hybrid algorithm SMPHA is evaluated in terms of the time taken to train the ML model in edge and cloud. In this experiment Raspberry Pi is used to train the SMPHA model with 196,400 rows, that is, input data sample size and takes around 1,710,000 ms (approximately 28.5 min). The same model when it is trained in Google Colab cloud environment, it takes 204,000 ms (approximately 3.4 min) as depicted in Table 3. The main purpose is to run the trained model on edge not to train the model at edge. So due to the lack of computing capability at the edge, it takes more time to train the model, but it can be ignored as it does not affect the purpose of the proposed model. Here, edge is introduced to obtain the task of computing from the cloud (i.e., offloading the task) by making the system more edge-oriented deployment. It can be accomplished rapidly as it requires only 14 s to download a trained SMPHA model from the cloud to the edge node with a size of 3101 kb as given in Table 3. The time to download varies according to the size of the trained model. So, from this process it can be inferred that downloading the trained model saves time when compared

Table 3 Comparison of model training time.

	Edge	Cloud
Model training time	28.4 min	3.4 min
Downloading time	Not applicable	14 s

to training the model at the edge. Through this in real time, deployment of the trained SMPHA model in edge is better compared to deployment in cloud services. Furthermore, network parameters like latency, throughput, bandwidth, and response time are adopted to measure the performance improvements in edge computing.

The performance metrices taken into account are latency, bandwidth, and response time [24]. The latency of an application is the product of two factors: computing latency and transmission latency. The time spent on data processing and transmission between end devices to cloud servers is termed as computing latency and transmission latency, respectively. The computational capacity of the system decides the computing latency as the network servers possess a considerable amount of capacity to make the data processing faster, whereas the sensors come with limited computing capacity. The latency in transmission is increased by the end devices and cloud servers. Bandwidth: As large number of sensors are deployed in IoT, data generated would be huge that consumes an intense range of bandwidth and leads to several problems such as delay in transmission and loss of packets. It becomes unacceptable for the data to be transferred directly to cloud servers without applying compression.

Therefore, data pre-processing and aggregation are needed for IoT gateways before redirecting them to remote cloud servers. Then, the issue to be confronted is to control the traffic flow by migrating data processing and aggregation tasks optimally to decrease the bandwidth needs of the end users while maintaining the data quality. Response time: The total response time is calculated by adding up transmission and processing time. The local deployment of the proposed model for controlling IoT-based irrigation are deployed on two modes: (i) Cloud mode: The developed SMPHA model is implemented in the cloud communicating with IoT sensors nodes directly to manage the irrigation process. The data are stored and processed at the cloud server itself where it uses Heroku platform. (ii) Edge mode – Raspberry Pi is deployed as an edge server that involves in processing of the SMPHA model controlling the IoT sensor nodes. Here, the data are stored and processed locally within the edge servers. This SMPHA model from both the edge and cloud does the job of controlling the actuators to initiate and quit the working of water flow motors.

Through this deployment in both the environments, performance of edge server and cloud server can be checked in terms of latency, throughput, bandwidth, and response time is shown in aforementioned graphs in Figs. 6–8. This performance metrics is not feasible to calculate while

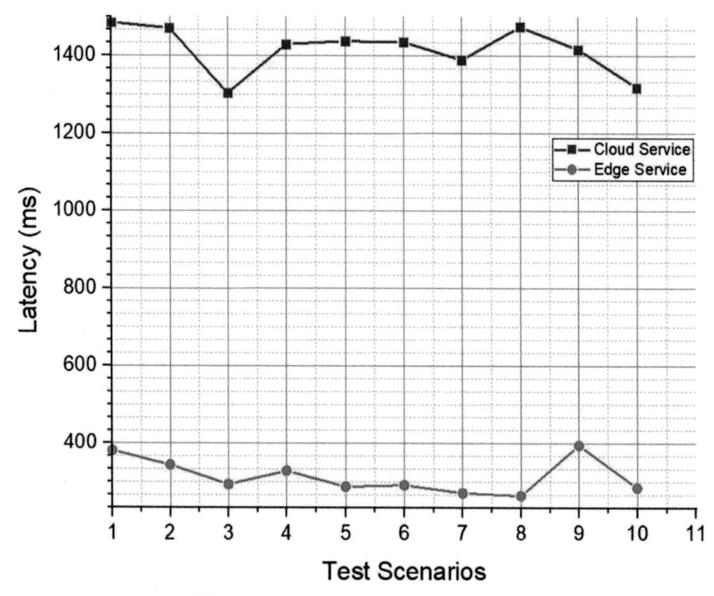

Fig. 6 Average latency with 10 test scenarios.

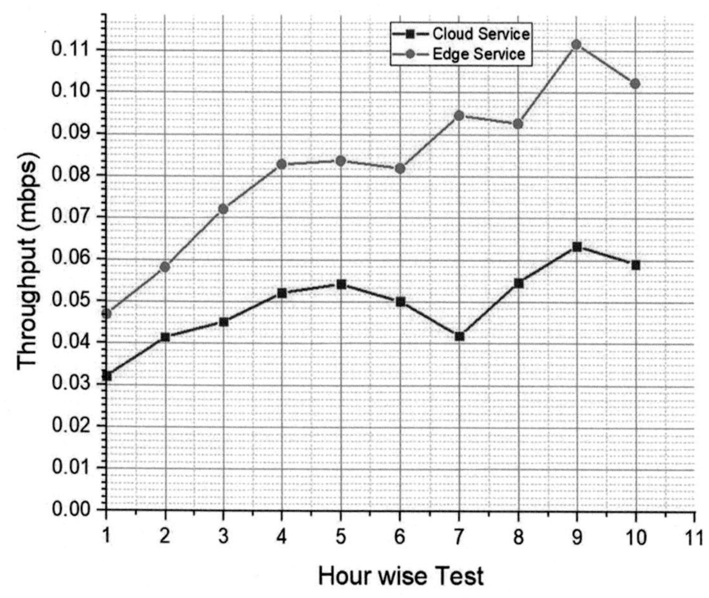

Fig 7 Average throughput value with 10 h test scenarios.

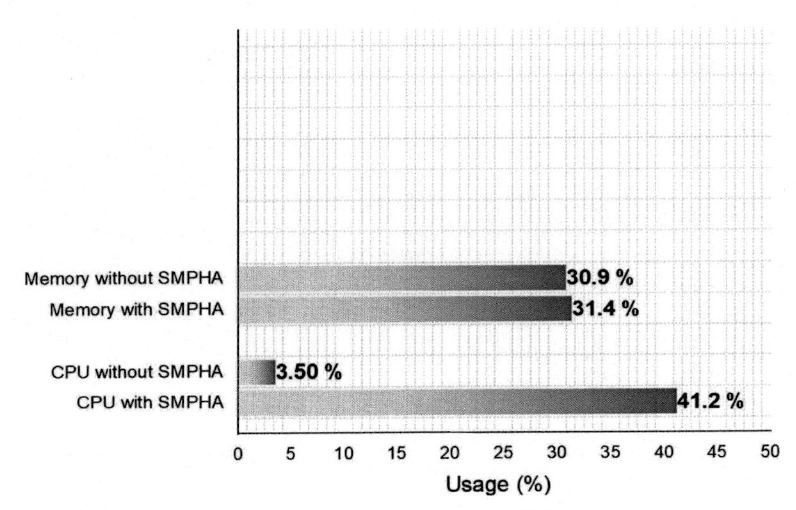

Fig. 8 CPU and memory utilization with and without SMPHA.

deploying in real time, so the aforementioned scenarios of two modes are virtually created by generating many request and response threads between the servers. This sampling, load test, and distributed testing are conducted through JMeter application and also verified with Wireshark in cloud servers. The test scenario is created here by data of sending and receiving sampling data between cloud to IoT sensors and between Edge to IoT sensors. The sampling data considered in this work refer to the approximate number of requests generated by Arduino to cloud and Arduino to Raspberry Pi that are calculated in real time. The test scenario is divided into 10 days of sampling data collected for each day. The evaluation results are depicted for latency and response times in 10 days perspective. In latency parameter, edge service has decreased by an average of 77.85% time compared to the with cloud. In the same manner, the response time of edge service is also decreased by 74.09% time compared to cloud service. In throughput calculation, sampling data are calculated for an hourly basis for the 10 h data in a day. From the hourly comparisons of throughput value, edge outperforms with 67.17% high Mbps usage. Through this analysis as shown in Table 4, it is evident that the proposed edge computing methodology deployed in Raspberry Pi or in local computers outperforms the cloud-oriented approach.

Finally, to illustrate the efficiency of resource management in edge computing, CPU and memory utilization are considered for the analysis as both factors rely on the service execution model and the computational needs of

Table 4 Performance metrics for cloud and edge services.

Performance metrics	Cloud service	Edge service
Throughput (Mbps)	0.04944	0.08265
Latency (ms)	1415.8	313.6
Response time (ms)	1519.6	393.8
Bandwidth (bps)	86	1365

the services being fired from off-loaders. Fig. 8 depicts the utilization of CPU and RAM on the Raspberry Pi acting as an edge node in two cases: with and without the deployment of SMPHA model on it. As shown in Fig. 8, the SMPHA model affects the CPU of the Raspberry Pi node significantly as it consumed around 41.2% of the CPU compared to only 3.5% when it does not host the SMPHA model. However, the memory (RAM) utilization in both the cases (with and without deployment of an SMPHA model) is nearly the same which is around 31%. Comparatively RAM utilization does not have much difference in with and without SMPHA. It is worthwhile to note that, the CPU utilization is still much lower than the 50% of total CPU capacity in Raspberry Pi. Therefore, it becomes feasible for adopting edge server implementation in the proposed irrigation system.

7.7 Conclusion

This article proposed a novel approach to edge-based irrigation system to facilitate decision-making on watering the plants on scheduled time. The proposed approach applying IoT with an edge computing framework enables the farming system to adapt to the changes in environmental conditions automatically and efficiently. The process of automatic irrigation regulates irrigation according to the live weather parameters for forecasting the irrigation process. Soil moisture prediction was performed using major regression algorithms that are again combined with k-means clustering for estimating the changes incurred in soil moisture prediction. These techniques were compared through metrics such as MAPE, MSE, speed, and power consumption from which XB + k-means was found to perform better. The XB + k-means algorithm was further used for the implementation of decision mechanism on the developed edge computing model. The proposed edge model saves the data communication cost and reduces the response time of IoT services. It can be deployed on existing devices on the network edges serving as edge nodes, thereby reducing the overall implementation cost of a large-scale IoT system. The edge-based approach was found to perform better than the cloud-based approach in terms of response time, latency, throughput, and bandwidth usage. Finally, the edge model was analyzed through CPU and memory usage while running with and without the algorithm. In both cases, the memory utilization is almost lower to total available resource of the edge device. From this, edge device can allocate its remaining resource for other computing services, which increases the efficiency of edge computing device. The number of end edge nodes can be increased according to the field area and then to check the potency of the system.

References

[1] S.K. Routray, A. Javali, L. Sharma, A.D. Ghosh, A. Sahoo, Internet of things based precision agriculture for developing countries, in: Proceedings of the 2019 International Conference on Smart Systems and Inventive Technology (ICSSIT), Tirunelveli, India, 27–29 November 2019, 2019, pp. 1064–1068.

[2] K. Perakis, F. Lampathaki, K. Nikas, Y. Georgiou, O. Marko, J. Maselyne, CYBELE–Fostering Precision Agriculture & Livestock Farming through Secure Access to large-scale HPC enabled virtual industrial experimentation environments fostering scalable big data analytics, Comput. Netw. 168 (2020) 107035.

[3] J. Poveda, Insect frass in the development of sustainable agriculture. A review, Agron. Sustain. Dev. 41 (2021) 1–10.

[4] P.C. Nagajyoti, K.D. Lee, T.V.M. Sreekanth, Heavy metals, occurrence and toxicity for plants: a review, Environ. Chem. Lett. 8 (2010) 199–216.

[5] P.K. Rai, S.S. Lee, M. Zhang, Y.F. Tsang, K.-H. Kim, Heavy metals in food crops: health risks, fate, mechanisms, and management, Environ. Int. 125 (2019) 365–385.

[6] L.S. Keith, D.W. Wohlers, D.B. Moffett, Z.A. Rosemond, ATSDR evaluation of potential for human exposure to tungsten, Toxicol. Ind. Health 23 (2007) 309–345.

[7] P.K. Rai, K.-H. Kim, S.S. Lee, J.-H. Lee, Molecular mechanisms in phytoremediation of environmental contaminants and prospects of engineered transgenic plants/microbes, Sci. Total Environ. 705 (2020) 135858.

[8] G. Sandeep, K.R. Vijayalatha, T. Anitha, Heavy metals and its impact in vegetable crops, Int. J. Chem. Stud. 7 (2019) 1612–1621.

[9] P.-I.K. Chukwuemeka, N.U. Hephzibah, Potential health risk from heavy metals via consumption of leafy vegetables in the vicinity of Warri refining and petrochemical company, Delta state, Nigeria, Ann. Biol. Sci. 6 (2018) 30–37.

[10] Y. Gao, P. Zhou, L. Mao, Y. Zhi, W. Shi, Assessment of effects of heavy metals combined pollution on soil enzyme activities and microbial community structure: modified ecological dose–response model and PCR-RAPD, Environ. Earth Sci. 60 (2010) 603–612.

[11] S. Tiwari, C. Lata, Heavy metal stress, signaling, and tolerance due to plant-associated microbes: An overview, Front. Plant Sci. 9 (2018) 452.

[12] H. Panchasara, N.H. Samrat, N. Islam, Greenhouse gas emissions trends and mitigation measures in Australian agriculture sector—a review, Agri 11 (2021) 85.

[13] S.R. Wild, K.C. Jones, Organic chemicals entering agricultural soils in sewage sludges: screening for their potential to transfer to crop plants and livestock, Sci. Total Environ. 119 (1992) 85–119.

[14] P.K. Rai, Impacts of particulate matter pollution on plants: implications for environmental biomonitoring, Ecotoxicol. Environ. Saf. 129 (2016) 120–136.
[15] M. Huang, Y. Zhu, Z. Li, B. Huang, N. Luo, C. Liu, G. Zeng, Compost as a soil amendment to remediate heavy metal-contaminated agricultural soil: mechanisms, efficacy, problems, and strategies, Water Air Soil Pollut. 227 (2016) 1–18.
[16] P.K. Rai, Biomagnetic Monitoring through Roadside Plants of an Indo-Burma Hot Spot Region, Elsevier, London, UK, 2016.
[17] R. Li, H. Wu, J. Ding, W. Fu, L. Gan, Y. Li, Mercury pollution in vegetables, grains and soils from areas surrounding coal-fired power plants, Sci. Rep. 7 (2017) 1–9.
[18] V. Fernández, T. Eichert, Uptake of hydrophilic solutes through plant leaves: current state of knowledge and perspectives of foliar fertilization, CRC Crit. Rev. Plant. Sci. 28 (2009) 36–68.
[19] R. Cavicchioli, W.J. Ripple, K.N. Timmis, F. Azam, L.R. Bakken, M. Baylis, et al., Scientists' warning to humanity: microorganisms and climate change, Nat. Rev. Microbiol. 17 (9) (2019) 569–586, https://doi.org/10.1038/s41579-019-0222-5.
[20] N.T.L. Huong, Y.S. Bo, S. Fahad, Economic impact of climate change on agriculture using Ricardian approach: a case of Northwest Vietnam, J. Saudi Soc. Agric. Sci. 18 (4) (2019) 449–457, https://doi.org/10.1016/j.jssas.2018.02.006.
[21] R.K. Fagodiya, H. Pathak, A. Bhatia, N. Jain, A. Kumar, S.K. Malyan, Global warming impacts of nitrogen use in agriculture: An assessment for India since 1960, Carbon Manag. 11 (3) (2020) 291–301, https://doi.org/10.1080/17583004.2020.1752061.
[22] S. Sarkar, S. Chatterjee, S. Misra, Assessment of the suitability of fog computing in the context of internet of things, IEEE Trans. Cloud Comput. 6 (1) (2018) 46–59, https://doi.org/10.1109/TCC.2015.2485206.
[23] K. Saravanan, G. Julie, H. Robinson, Handbook of Research on Implementation and Deployment of IoT Projects in Smart Cities, IGI Global, Hershey, 2019, https://doi.org/10.4018/978-1-5225-9199-3.
[24] A. Baylis, Advances in precision farming technologies for crop protection, Outlooks Pest. Manag. 28 (4) (2017) 158–161, https://doi.org/10.1564/v28_aug_04.

Further reading

[25] C.F. Nicholson, E.C. Stephens, B. Kopainsky, P.K. Thornton, A.D. Jones, D. Parsons, J. Garrett, Food security outcomes in agricultural systems models: case examples and priority information needs, Agr. Syst. 188 (2021) 103030.
[26] C. Lopes, M. Herva, A. Franco-Uría, E. Roca, Inventory of heavy metal content in organic waste applied as fertilizer in agriculture: evaluating the risk of transfer into the food chain, Environ. Sci. Pollut. Res. 18 (2011) 918–939.
[27] M. Arora, B. Kiran, S. Rani, A. Rani, B. Kaur, N. Mittal, Heavy metal accumulation in vegetables irrigated with water from different sources, Food Chem. 111 (2008) 811–815.
[28] A.K. Meena, G.K. Mishra, P.K. Rai, C. Rajagopal, P.N. Nagar, Removal of heavy metal ions from aqueous solutions using carbon aerogel as an adsorbent, J. Hazard. Mater. 122 (2005) 161–170.
[29] P.K. Rai, Heavy metal phytoremediation from aquatic ecosystems with special reference to macrophytes, Crit. Rev. Environ. Sci. Technol. 39 (2009) 697–753.
[30] J.E. Gall, R.S. Boyd, N. Rajakaruna, Transfer of heavy metals through terrestrial food webs: a review, Environ. Monit. Assess. 187 (2015) 1–21.
[31] Z.J. Shen, Y.S. Chen, Z. Zhang, Heavy metals translocation and accumulation from the rhizosphere soils to the edible parts of the medicinal plant Fengdan (Paeonia ostii) grown on a metal mining area, China, Ecotoxicol. Environ. Saf. 143 (2017) 19–27.

[32] O. El Hamiani, H. El Khalil, C. Sirguey, A. Ouhammou, G. Bitton, C. Schwartz, A. Boularbah, Metal concentrations in plants from mining areas in South Morocco: health risks assessment of consumption of edible and aromatic plants, CLEAN Soil Air Water 43 (2015) 399–407.
[33] S. Bolan, A. Kunhikrishnan, B. Seshadri, G. Choppala, R. Naidu, N.S. Bolan, Y.S. Ok, M. Zhang, C.G. Li, F. Li, Sources, distribution, bioavailability, toxicity, and risk assessment of heavy metal (loid) s in complementary medicines, Environ. Int. 108 (2017) 103–118.
[34] S.W. Kim, Y.E. Chae, J.M. Moon, D.K. Kim, R.X. Cui, G. An, S.W. Jeong, Y.J. An, In situ evaluation of crop productivity and bioaccumulation of heavy metals in Paddy soils after remediation of metal-contaminated soils, J. Agric. Food Chem. 65 (2017) 1239–1246.
[35] S. Kohzadi, B. Shahmoradi, E. Ghaderi, H. Loqmani, A. Maleki, Concentration, source, and potential human health risk of heavy metals in the commonly consumed medicinal plants, Biol. Trace Elem. Res. 187 (2019) 41–50.
[36] F. Li, W. Shi, Z. Jin, H. Wu, G.D. Sheng, Excessive uptake of heavy metals by greenhouse vegetables, J. Geochem. Explor. 173 (2017) 76–84.
[37] L. Yu, G. Xin, W. Gang, Q. Zhang, S. Qiong, X. Guoju, Heavy metal contamination and source in arid agricultural soil in Central Gansu Province, China, J. Environ. Sci. 20 (2008) 607–612.
[38] A.K. Chopra, C. Pathak, G. Prasad, Scenario of heavy metal contamination in agricultural soil and its management, J. Appl. Nat. Sci. 1 (2009) 99–108.
[39] W. Feng, Z. Guo, X. Xiao, C. Peng, L. Shi, H. Ran, W. Xu, A dynamic model to evaluate the critical loads of heavy metals in agricultural soil, Ecotoxicol. Environ. Saf. 197 (2020) 110607.
[40] J. Wu, J. Li, Y. Teng, H. Chen, Y. Wang, A partition computing-based positive matrix factorization (PC-PMF) approach for the source apportionment of agricultural soil heavy metal contents and associated health risks, J. Hazard. Mater. 388 (2020) 121766.
[41] M. Shahid, C. Dumat, S. Khalid, E. Schreck, T. Xiong, N.K. Niazi, Foliar heavy metal uptake, toxicity and detoxification in plants: a comparison of foliar and root metal uptake, J. Hazard. Mater. 325 (2017) 36–58.
[42] R. Lal, Adaptation and mitigation of climate change by improving agriculture in India, in: S. SherazMahdi (Ed.), Climate Change and Agriculture in India: Impact and Adaptation, Springer International Publishing, Cham, 2019, pp. 217–227, https://doi.org/10.1007/978-3-319-90086-5_17.
[43] D. Mulla, R. Khosla, Historical Evolution and Recent Advances in Precision Farming. Soil-Specific Farming Precision Agriculture, CRC Press, Boca Raton, 2015, https://doi.org/10.1201/b18759-2.
[44] L. Dutta, T.K. Basu, Extraction and optimization of leaves images of mango tree and classification using ANN, IJRAET 1 (3) (2013) 46–51.
[45] T. Kawai, H. Mineno, Evaluation environment using edge computing for artificial intelligence-based irrigation system, in: 2020 16th International Conference on Mobility, Sensing and Networking (MSN), IEEE, Tokyo, Japan, 2020, pp. 214–219, https://doi.org/10.1109/MSN50589.2020.00046.
[46] M.S. Munir, I.S. Bajwa, A. Ashraf, W. Anwar, R. Rashid, Intelligent and smart irrigation system using edge computing and IoT, Complexity. 2021 (2021) 1–16, https://doi.org/10.1155/2021/6691571.
[47] C.M. Angelopoulos, G. Filios, S. Nikoletseas, T.P. Raptis, Keeping data at the edge of smart irrigation networks: a case study in strawberry greenhouses, Comput. Netw. 167 (2020) 107039, https://doi.org/10.1016/j.comnet.2019.107039.
[48] M. Satyanarayanan, The emergence of edge computing, Computer. 50 (1) (2017) 30–39, https://doi.org/10.1109/MC.2017.9.

[49] W. Shi, S. Dustdar, The promise of edge computing, Computer. 49 (5) (2016) 78–81, https://doi.org/10.1109/MC.2016.145.
[50] P.L. Ramirez Izolan, F. Diniz Rossi, R. Hohemberger, M.P. Konzen, R.G. da Cunha, L.R. Saquette, et al., Low-cost fog computing platform for soil moisture management, in: 2020 International Conference on Information Networking (ICOIN), IEEE, Barcelona, Spain, 2020, pp. 499–504, https://doi.org/10.1109/ICOIN48656.2020.9016572.
[51] F. Ferrandez-Pastor, J. Garcia-Chamizo, M. Nieto-Hidalgo, J. Mora-Pascual, J. Mora-Martínez, Developing ubiquitous sensor network platform using internet of things: application in precision agriculture, Sensors. 16 (7) (2016) 1141, https://doi.org/10.3390/s16071141.
[52] X. Xu, X. Liu, Z. Xu, F. Dai, X. Zhang, L. Qi, Trust-oriented IoT service placement for smart cities in edge computing, IEEE Internet Things J. 7 (5) (2020) 4084–4091, https://doi.org/10.1109/JIOT.2019.2959124.
[53] X. Wu, M. Liu, In-situ soil moisture sensing: measurement scheduling and estimation using compressive sensing, in: 2012 ACM/IEEE 11th International Conference on Information Processing in Sensor Networks (IPSN), IEEE, Beijing, China, 2012, pp. 1–11, https://doi.org/10.1145/2185677.2185679.
[54] T. Kameoka, K. Nishioka, Y. Motonaga, Y. Kimura, A. Hashimoto, Watanabe N. smart sensing in a vineyard for advanced viticultural management, in: Proceedings of the 2014 International Workshop on Web Intelligence and Smart Sensing, Saint Etienne France, 2014, pp. 1–4, https://doi.org/10.1145/2637064.2637091.
[55] K. Cagri Serdaroglu, C. Onel, S. Baydere, IoT-based smart plant irrigation system with enhanced learning, in: 2020 IEEE Computing, Communications and IoT Applications (ComComAp.), IEEE, Beijing, China, 2020, pp. 1–6, https://doi.org/10.1109/ComComAp51192.2020.9398892.
[56] J. Kwok, Y. Sun, A smart IoT-based irrigation system with automated plant recognition using deep learning, in: Proceedings of the 10th International Conference on Computer Modeling and Simulation - ICCMS2018, ACM Press, Sydney, Australia, 2018, pp. 87–91, https://doi.org/10.1145/3177457.3177506.
[57] A. Goldstein, L. Fink, A. Meitin, S. Bohadana, O. Lutenberg, G. Ravid, Applying machine learning on sensor data for irrigation recommendations: revealing the agronomist's tacit knowledge, Precision Agricult. 19 (3) (2018) 421–444, https://doi.org/10.1007/s11119-017-9527-4.
[58] A. Vij, S. Vijendra, A. Jain, S. Bajaj, A. Bassi, A. Sharma, IoT and machine learning approaches for automation of farm irrigation system, Proc. Comput. Sci. 167 (2020) 1250–1257, https://doi.org/10.1016/j.procs.2020.03.440.
[59] H. Krishnan, R. Scholar, MongoDB – a comparison with NoSQL databases, Int. J. Sci. Eng. Res. 7 (5) (2016) 1035–1037.
[60] T. Ojha, S. Misra, N.S. Raghuwanshi, Wireless sensor networks for agriculture: the state-of-the-art in practice and future challenges, Comput Electr Agricult. 118 (2015) 66–84, https://doi.org/10.1016/j.compag.2015.08.011.
[61] J. Gutierrez, J.F. Villa-Medina, A. Nieto-Garibay, M.A. Porta-Gandara, Automated irrigation system using a wireless sensor network and GPRS module, IEEE Trans. Instrum. Meas. 63 (1) (2014) 166–176, https://doi.org/10.1109/TIM.2013.2276487.
[62] S. Chanthakit, P. Keeratiwintakorn, C. Rattanapoka, An IoT system design with real time stream processing and data flow integration, in: In: 2019 Research, Invention, and Innovation Congress (RI2C.), IEEE, Bangkok, Thailand, 2019, pp. 1–5, https://doi.org/10.1109/RI2C48728.2019.8999968.
[63] H. Lv, S. Wang, Design and Application of IoT Microservices Based on Seneca, DEStech Transactions on Computer Science and Engineering, (icte.), USA, 2016, https://doi.org/10.12783/dtcse/icte2016/4814.

[64] B.-H. Lee, E.K. Dewi, M.F. Wajdi, Data security in cloud computing using AES under HEROKU cloud, in: 2018 27th Wireless and Optical Communication Conference (WOCC), IEEE, Hualien, 2018, pp. 1–5, https://doi.org/10.1109/WOCC.2018.8372705.
[65] M.A. Lopez Pena, F.I. Munoz, SAT-IoT: An architectural model for a high-performance fog/edge/cloud IoT platform, in: 2019 IEEE 5th World Forum on Internet of Things (WF-IoT.), IEEE, Limerick, Ireland, 2019, pp. 633–638, https://doi.org/10.1109/WF-IoT.2019.8767282.
[66] Weather API. Retrieved from https://openweathermap.org/api.
[67] D. Gislason, Zigbee Wireless Networking, first ed., Elsevier Publisher, Newnes, London, 2008.
[68] K. Tanabe, Y. Tanabe, M. Hagiya, Model-based testing for MQTT applications, in: M. Virvou, H. Nakagawa, L.C. Jain (Eds.), Knowledge-Based Software Engineering: 2020, Springer International Publishing, Cham, 2020, pp. 47–59, https://doi.org/10.1007/978-3-030-53949-8_5.
[69] L. Babun, K. Denney, Z.B. Celik, P. McDaniel, A.S. Uluagac, A survey on IoT platforms: communication, security, and privacy perspectives, Comput. Netw. 192 (2021) 108040, https://doi.org/10.1016/j.comnet.2021.108040.
[70] K. Rastogi, D. Lohani, Edge computing-based internet of things framework for indoor occupancy estimation, Int. J. Ambient Comput. Intell. 11 (4) (2020) 16–37, https://doi.org/10.4018/978-1-6684-5700-9.ch031.
[71] S. Premkumar, A.N. Sigappi, Functional framework for edge-based agricultural system, in: AI, Edge and IoT-Based Smart Agriculture, first ed., Academic Press, Elsevier, USA, 2021, pp. 71–100, https://doi.org/10.1016/B978-0-12-823694-9.00029-3.
[72] J. Phani Kumar, P. Paramaguru, T. Arumugam, N. Manikanda Boopathi, K. Venkatesan, Genetic divergence among Ramnad mundu chilli (Capsicum annuum L.) genotypes for yield and quality, Electr. J. Plant Breeding. 12 (1) (2021) 228–234.
[73] A. Goap, D. Sharma, A.K. Shukla, K.C. Rama, An IoT-based smart irrigation management system using machine learning and open source technologies, Comput Electronic Agricult. 155 (2018) 41–49, https://doi.org/10.1016/j.compag.2018.09.040.
[74] M.S. Aslanpour, S.S. Gill, A.N. Toosi, Performance evaluation metrics for cloud, fog and edge computing: a review, taxonomy, benchmarks and standards for future research, Internet Things. 12 (2020) 100273, https://doi.org/10.1016/j.iot.2020.100273.
[75] A. Sunardi, Suharjito MVC architecture: a comparative study between Laravel framework and slim framework in freelancer project monitoring system web based, Proc. Comput. Sci. 157 (2019) 134–141, https://doi.org/10.1016/j.procs.2019.08.150.
[76] R. Shimonski, The Wireshark Field Guide, first ed., Syngress Press, Elsevier, New York, 2013, https://doi.org/10.1016/B978-0-12-410413-6.00001-2.

About the authors

Dr. Preetha Evangeline David is currently working as an Associate Professor and Head of the Department in the Department of Artificial Intelligence and Machine Learning at Chennai Institute of Technology, Chennai, India. She holds a PhD from Anna University, Chennai in the area of Cloud Computing. She has published many research papers and Patents focusing on Artificial Intelligence, Digital Twin Technology, High Performance Computing, Computational Intelligence and Data Structures. She is currently working on Multi-disciplinary areas in collaboration with other technologies to solve socially relevant challenges and provide solutions to human problems.

Dr. P. Anandhakumar is a professor in the Department of Information Technology at Anna University, Chennai. He has completed his doctorate in the year 2006 from Anna University. He has produced 17 PhD's in the field of Image Processing, Cloud Computing, Multimedia technology and Machine Learning. His ongoing research lies in the field of Digital Twin Technology, Machine Learning and Artificial Intelligence. He has published more than 150 papers indexed in SCI, SCOPUS, WOS etc.

Pethuru Raj Chelliah (PhD) works as the chief architect at the Site Reliability Engineering Center of Excellence, Reliance Jio Infocomm Ltd. (RJIL), Bangalore. Previously, he worked as a cloud infrastructure architect at the IBM Global Cloud Center of Excellence, IBM India, Bangalore, for four years. He also had an extended stint as a TOGAF-certified enterprise architecture consultant in Wipro Consulting services division and as a lead architect in the corporate research division of Robert Bosch, Bangalore. He has more than 17 years of IT industry experience. Shreyash Naithani is currently a site reliability engineer at Microsoft R&D. Prior to Microsoft, he worked with both start-ups and mid-level companies. He completed his PG Diploma from the Centre for Development of Advanced Computing, Bengaluru, India, and is a computer science graduate from Punjab Technical University, India. In a short span of time, he has had the opportunity to work as a DevOps engineer with Python/C#, and as a tools developer, site/service reliability engineer, and Unix system administrator. During his leisure time, he loves to travel and binge watch series. Shailender Singh is a principal site reliability engineer and a solution architect with around 11 year's IT experience who holds two master's degrees in IT and computer application. He has worked as a C developer on the Linux platform. He had exposure to almost all infrastructure technologies from hybrid to cloud-hosted environments. In the past, he has worked with companies including Mckinsey, HP, HCL, Revionics and Avalara and these days he tends to use AWS, K8s, Terraform, Packer, Jenkins, Ansible, and OpenShift.

CHAPTER TEN

An automatic path navigation for visually challenged people using deep Q learning

S. Muthurajkumar, B. Rahul, L.S. Sanjay Kumar, and E. Gokkul
Department of Computer Technology, Madras Institute of Technology (MIT) Campus, Anna University, Chrompet, Chennai, India

Contents

Abstract

There is an increased need for automation in the ever-advancing technological world. Automation reduces time, money, labor, while also reducing manual errors, giving you more time to concentrate on other work. Manual tasks are hectic and boring, sometimes they are dangerous considering the work. Navigation plays a vital role in many industries, some of them are vehicles, drones, transportation, and there are certain cases where navigation might be dangerous like deep forest, underground, underwater, firefighting environments. Visually challenged people have a difficult time navigating unfamiliar places. So, there is a necessity for an automatic navigation system to overcome these situations. Hence, in this research work, the automatic path navigation simulation using deep Q-learning is proposed. The proposed path navigation system when assigned its destination automatically finds the shortest route and avoids the obstacle in order to reach its destination.

Advances in Computers, Volume 132
ISSN 0065-2458
https://doi.org/10.1016/bs.adcom.2023.07.005

1. Introduction

A path navigation system using Deep Q learning is presented in our research work. Initially an environment is created with the help of Kivy software, then an object will be created with three sensors. The object will have to avoid obstacles that are created dynamically by the user and reach its destination by taking the shortest path. Here, we will be using Deep Q Learning algorithm (Reinforcement Learning) wherein the object will be trained based on reward and penalty mechanism. The main aim of our research work is to make an object automatically navigate to its destination by avoiding the obstacles.

Our first objective is to build an environment with an object and three sensors (one right sensor, one left sensor and one straight sensor). Second, to build an obstacle creator and assign a function to the object (i.e.) the object must turn left or right in order to avoid the obstacle. Finally, to implement deep Q learning algorithm to train the object in order to avoid the obstacle and find the shortest path to reach its destination.

2. Literature survey

Arvind et al. [1] have developed a static obstacle detection using reinforcement learning for autonomous vehicle navigation in a simulated environment. MLP-NN predicted the next action based on vehicle acceleration, heading angle, distance measure from the ultrasonic sensor.

Babu et al. [2] have implemented a path and motion planning for a robot were used to make it autonomous in unknown environment. These were achieved using image processing and reinforcement techniques using Q Learning. The authors calculated the hottest path from current state to goal state by analyzing the environment through captured images.

Beakcheol et al. [3] Q-learning algorithms were off policy reinforcement learning algorithms that tried performing the most profitable action given the current state. They have covered all variants of Q-learning algorithms, which are a representative algorithm under reinforcement learning. They distinctively categorized Q-learning algorithms into single-agent and multi-agent and described them thoroughly.

Lan et al. [4] combined deep Q learning with learning replay and heuristic knowledge for path detection and obstacle avoidance of intelligent robots.

Jiangdong et al. [5] developed a DRL enabled highway overtaking driving policy for autonomous vehicles. They proposed a decision-making strategy and it was evaluated and estimated to be adaptive to other complicated scenarios. First, the studied driving environment was founded on the highway, wherein an ego vehicle aims to run through a particular driving scenario efficiently and safely. Finally, the performance of the proposed control framework was discussed via executing a series of simulation experiments.

Hyansu et al. [6] developed with information and strategy around reinforcement learning for multi-robot navigation algorithms where each robot can be considered as a dynamic obstacle or cooperative robot depending on the situation. Each robot in the system can perform independent actions and simultaneously collaborate with each other depending on the given mission. After the selected action, the relationship with the target was evaluated, and rewards or penalty was given to each robot to learn.

Lemos et al. [7] were aimed to present the results of an assessment of adherence to the Deep Q-learning algorithm, applied to a vehicular navigation robot. The robot's job was to transport parts through an environment, for their purpose, a decision system was built based on the Deep Q-learning algorithm, with the aid of an artificial neural network that received data from the sensors as input and allowed autonomous navigation in an environment. For the experiments, the mobile robot-maintained communication via the network with other robotic components present in the environment through the MQTT protocol.

Chao et al. [8] have proposed a Deep Reinforcement Learning (DRL) approach for UAV path planning based on the global situation information. They have chosen the STAGE Scenario software to provide the simulation environment where a situation assessment model was developed with consideration of the UAV survival probability under enemy radar detection and missile attack.

There are many works related to reinforcement learning algorithms and more method in [9–13].

Runnan et al. [14] developed were aiming at the path planning of mobile robots in UDE, a continuous dynamic simulation environment was built in their work. Based on DQN, a reward function with reward weight was designed, and the influence of reward weight has been analyzed experimentally. Moreover, the abnormal rewards caused by the relative motion between obstacles and robots were analyzed and solved by adding a reward modifier to DQN. The comparative experiment among RMDQN,

RMDDQN, dueling RMDQN, and dueling RMDDQN was done, and it turns out that the result of RMDDQN is the best.

Nguyen et al. [15] were aimed to illustrate how the Omni robot performs navigation using model-free deep Q learning to navigate in unpredicted environments. It also explained how to obtain the policy when such a model was unknown in advance by using a virtual environment to conduct in simulation. There are many works related to machine learning algorithms and more method in [16–20].

3. Proposed work

This architecture diagram clearly depicts our goal and outline structure of our research work.

3.1 Create object/vehicle

This module creates the object required, that is the vehicle using Kivy with any desired shape and size.

3.2 Sensors creation

This module creates the three sensors, the left sensor, the right sensor and the straight. The main purpose of these sensors is to detect the obstacles, if there are any.

3.3 Connecting Objects and Sensors

This module connects the 4 objects into a single entity, that is it connects the object with three sensors.

3.4 Obstacle map creation

In this module, we create an environment/map for the object to navigate from source to destination. We have also added clear (deletes the obstacle), save (saves the obstacle design) and load (loads the previously saved obstacle design) options.

As shown in Fig. 1, the architecture diagram has six components,

3.5 DQL algorithm implementation

This module deals with the DQL Algorithm which is used by the object to learn to avoid obstacles. If the object detects any obstacle from any of the three sensors it turns to the opposite direction by 20 degrees and moves

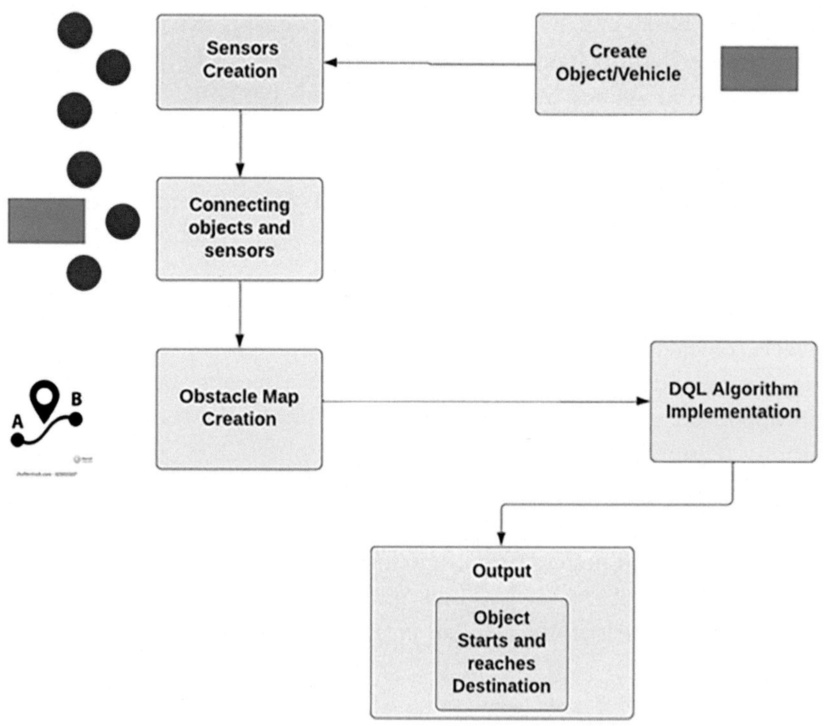

Fig. 1 Architecture diagram.

towards its destination. Here reinforcement learning method is used along with deep Q learning where they will be awarded with a reward for every step the move closer to the destination and a penalty will be given when it moves away from the destination and aim of it is to maximize its reward points.

3.6 Object starts and reaches destination

This module deals with the object moving from its initial point by dodging all the obstacles and by finding the shortest route to reach to the designated destinations.

The goal of this simulation is to build an environment with a complex path, an object with three sensors which uses Deep Q-Learning to train the object and to assign a destination for the object to reach.

Anaconda is what we will use to install PyTorch and Kivy. It is a free and open-source distribution of Python which offers an easy way to install packages.

We will build this 2D map inside a Kivy webapp. Kivy is a free and open-source Python framework with a user interface inside which you can build your games or apps. It will be the container of the whole environment.

PyTorch is the AI framework used to build our Deep Q-Network. PyTorch is great to work with and powerful. It has dynamic graphs which allow fast computations of the gradient of complex functions, needed to train the model.

We create the environment and we use Kivy WebApp to create 4 Kivy objects, a rectangle shape representing the object and three sensors to detect any obstacle and to navigate to the destination. We set our object to go from the upper left corner of the map, to the bottom right corner. Create 3 buttons: Clear, Load and Save.

We build a system to draw different obstacles in the environment. We assign functions to the objects to make it go through any path we create from the start to the end point. Assign function to Clear button.

Using Deep Q-Learning we build and train our object to navigate its way avoiding any obstacles to its destination. Assign functions for Load and save buttons.

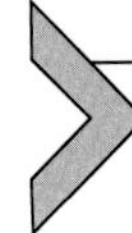

4. Implementation and algorithm

Step 1: Importing the libraries and the Kivy packages

Step 2: Initializing variables to keep the last point in memory when we draw the sand on the map, the total number of points in the last drawing, the length of the last drawing

Step 3: Create the brain of our AI, list of actions and the reward variable

- **i.** Four inputs, three actions, gamma = 0.9.
- **ii.** action = 0 => no rotation, action = 1 => rotate 20 degrees, action = 2 => rotate −20 degrees.
- **iii.** The reward received after reaching a new state.

Step 4: Initializing the map

- **i.** Sand is an array that has as many cells as our graphic interface has pixels. Each cell has a one if there is sand, 0 otherwise.
- **ii.** Building x-coordinate and y-coordinate of the goal.
- **iii.** Initializing the sand array with only zeros.
- **iv.** The goal to reach is at the upper left of the map. (The x-coordinate and y-coordinate).
- **v.** Initializing the last distance from the object to the goal.

Step 5: Creating the object class

i. Initializing the angle of the object.
ii. Initializing the last rotation of the object.
iii. Initializing the x-coordinate and y-coordinate of the velocity vector and the velocity vector.
iv. Initializing the x-coordinate and y-coordinate of all three sensors and their respective sensor vectors.

Step 6: Updating the position of the object according to its last position and velocity

i. Getting the rotation of the object.
ii. Updating the angle and the position of sensors 1, 2 and 3.
iii. Updating the signal received by sensors 1, 2 and 3. (Density of sand around sensor 1, 2, 3.)
iv. If any sensor is out of the map (the object is facing one edge of the map) that sensor detects full sand.
v. Update sensors 1, 2 and 3.

Step 7: Creating the game class

i. Getting the object and the sensors 1, 2 and 3 from our kivy file.
ii. Starting the object when we launch the application.
iii. The object will start at the center of the map.
iv. The object will start to go horizontally to the right with a speed of 6.

Step 8: Update function that updates everything that needs to be updated at each discrete time when reaching a new state (getting new signals from the sensors)

i. Specifying the global variables.
ii. Store width and height of the map (horizontal edge and vertical edge).
iii. Storing the difference of x-coordinates and of y-coordinates between the goal and the object.
iv. Initializing the direction of the object with respect to the goal (if the object is heading perfectly towards the goal, then orientation = 0)
v. Initializing our input state vector, composed of the orientation plus the three signals received by the three sensors.
vi. Updating the weights of the neural network in our ai and playing a new action
vii. Converting the action played (0, 1 or 2) into the rotation angle (0, 20 or −20 angles)
viii. Moving the object according to this last rotation angle
ix. Getting the new distance between the object and the goal right after the object moved
x. Updating the positions of the three sensors 1, 2 and 3 right after the object moved.

Step 9: Assigning reward system

i. If the object is on the sand, it is slowed down (speed=1) and reward=−1.
ii. Otherwise, it gets a bad reward of −0.2.
iii. However, if it is getting closer to the goal it still gets a slightly positive reward of 0.1.
iv. If the object is on any edge of the frame (top, right, bottom, left), it comes back 10 pixels away from the edge and it gets a bad reward of −1.
v. When the object reaches its goal, the goal becomes the bottom right corner of the map and vice versa (updating of the x and y coordinate of the goal).
vi. Updating the last distance from the object to the goal.

Step 10: Painting for graphic interface

i. Putting some sand when we do a left click.
ii. Put some sand when we move the mouse while pressing left.

Step 11: API and switches interface

i. Building the app.
ii. Creating the clear, save and load buttons.
iii. Running the app.

5. Results and output

The environment with objects is created, the object is set to go from the upper left corner of the map to the bottom right corner.

Fig. 2 shows the object learning the environment, and it reaches the starting point back from the destination. This process continues repeatedly as the object learns the environment better and better. Finally, the object follows the best path through the obstacle.

Fig. 3 shows Performance analysis is done for the environment in Obstacle-1. First, we can notice a lot of hits at the obstacle by the object. Once it learns the environment, we can notice a drastic change in the number of hits per journey.

Fig. 4 shows Performance analysis is done for the environment in Obstacle-2. First, we can notice a lot of hits at the obstacle by the object. Once it learns the environment, we can notice a drastic change in the number of hits per journey.

Fig. 5 shows Performance analysis is done for the environment in Obstacle-3. First, we can notice a lot of hits at the obstacle by the object.

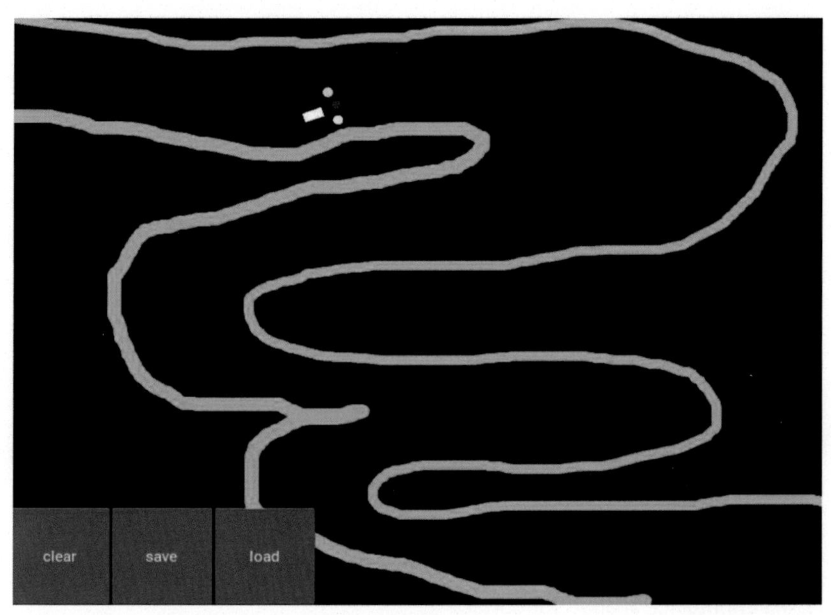

Fig. 2 Obstacle 3 (object reaching its destination).

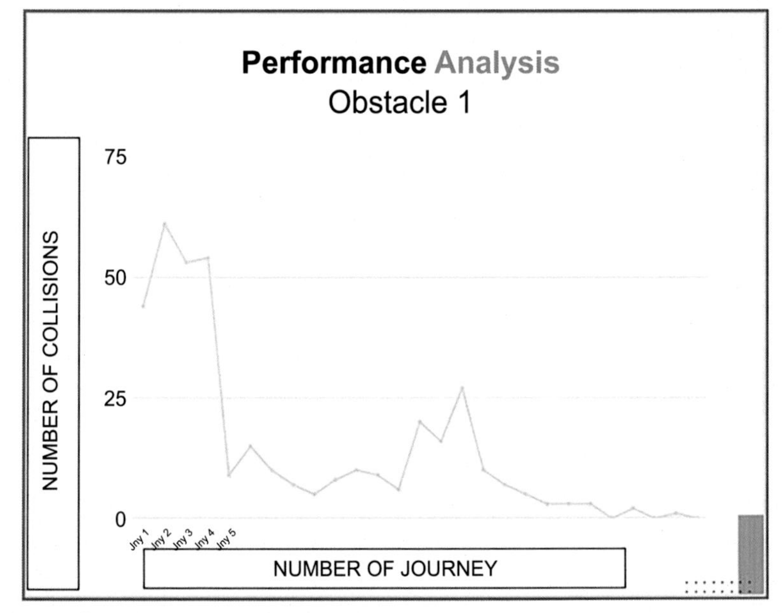

Fig. 3 Performance analysis for obstacle 1.

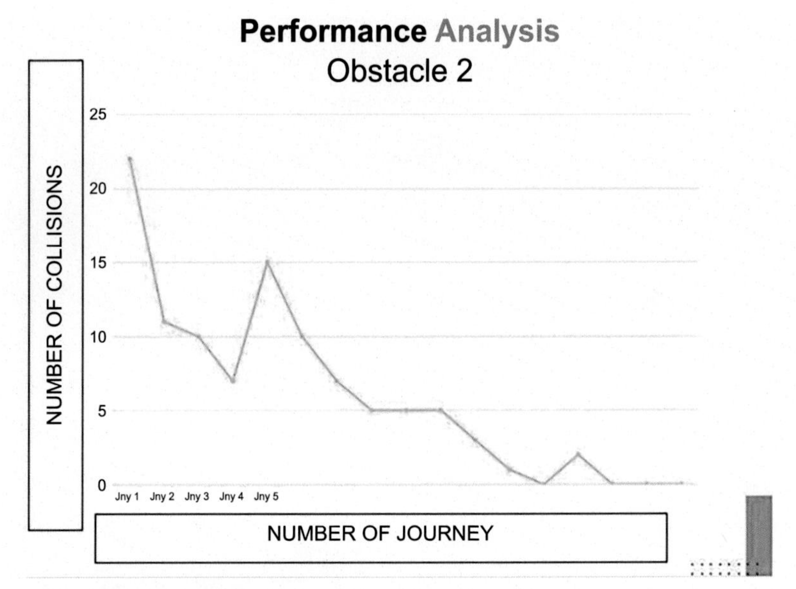

Fig. 4 Performance analysis for obstacle 2.

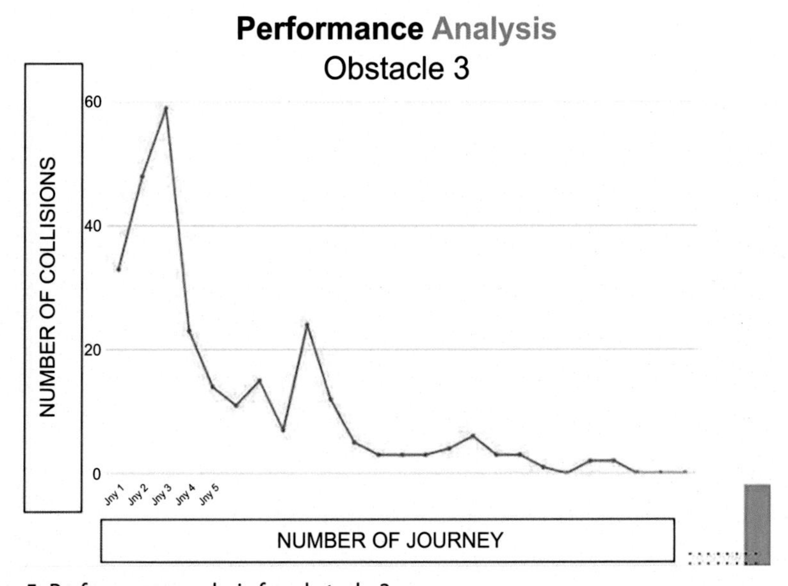

Fig. 5 Performance analysis for obstacle 3.

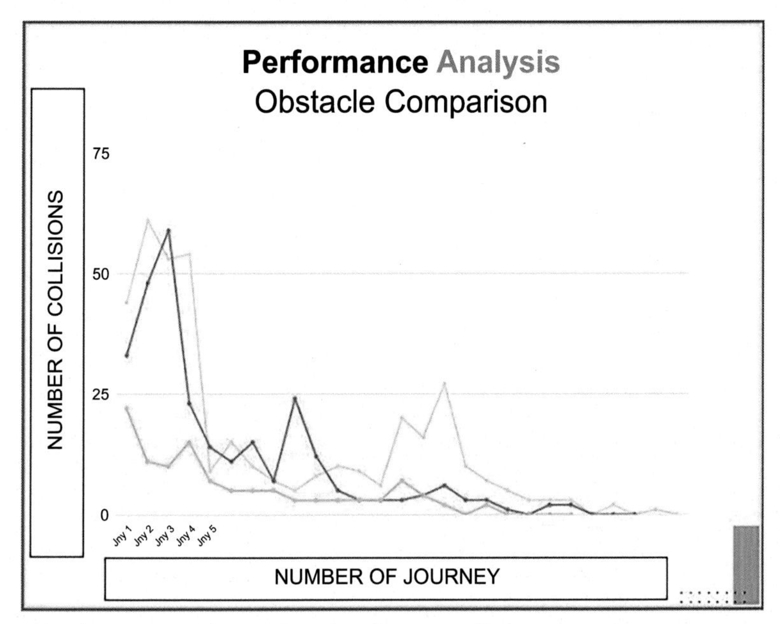

Fig. 6 Performance analysis comparison between all obstacle environments.

Once it learns the environment, we can notice a drastic change in the number of hits per journey.

Fig. 6 shows Performance analysis comparison is done for all obstacle environments. This diagram depicts the complexity of each path. More the complexity, the more the number of hits at the obstacle at the beginning.

6. Conclusion

Thus, we have developed a deep Q-learning based model trained in a virtual environment that is able to make decisions for navigation in an adaptive way. As inputs it took the information from the three sensors and its current orientation. As output, it decided the Q-values for each of the actions of going straight, turning left or turning right. As for the rewards, we punished it badly for hitting the sand, punished it slightly for going in the wrong direction and rewarded it slightly for going in the right direction.

Kivy was used to emulate the fire environment and PyTorch was used to communicate data and controls between the virtual environment and the deep learning model. The model was successfully able to navigate extreme fires based on its acquired knowledge and experience.

This research work serves as the foundation on which to build a deep learning framework that could identify objects within the environment and incorporating those objects into its decision-making process in order to successfully deliver safe, navigable routes to firefighters.

References

[1] C.S. Arvind, Autonomous vehicle for obstacle detection and avoidance using reinforcement learning, International Conference on Soft Computing for Problem Solving (2019).

[2] V.M. Babu, U. Vamshi Krishna, S.K. Shahensha, An autonomous path finding robot using Q-learning, in: 2016 10th International Conference on Intelligent Systems and Control (ISCO), IEEE, 2016, pp. 1–6.

[3] Beakcheol J., Myeonghwi K., Gaspard H., Kim J.W., Q-learning algorithms: A comprehensive classification and applications, IEEE Access 7 (2019) 133653–133667.

[4] Lan J., Hongyun H., Ding Z., Path planning for intelligent robots based on deep Q-learning with experience replay and heuristic knowledge, IEEE/CAA Journal of Automatica Sinica 7 (4) (2019) 1179–1189.

[5] Jiangdong L., Teng L., Xiaolin T., Xingyu M., Bing H., Cao D., Decision-making strategy on highway for autonomous vehicles using deep reinforcement learning, IEEE Access 8 (2020) 177804–177814.

[6] Hyansu B., Gidong K., Jonguk K., Dianwei Q., Lee S., Multi-robot path planning method using reinforcement learning, Applied Sciences 9 (15) (2019) 3057.

[7] M. Lemos, A. de Souza, R. de Lira, C. de Freitas, V. da Silva, Robot navigation through the deep Q-learning algorithm, International Journal of Advanced Engineering Research and Science 8 (2) (2021).

[8] Chao Y., Xiaojia X., Wang C., Towards real-time path planning through deep reinforcement learning for a UAV in dynamic environments, Journal of Intelligent & Robotic Systems 98 (2020) 297–309.

[9] M.R. Lemos, A.V. Rodrigues de Souza, R. Souza de Lira, C.A. Oliveira de Freitas, V. João da Silva, V. Ferreira de Lucena, Robot training and navigation through the deep Q-Learning algorithm, in: 2021 IEEE International Conference on Consumer Electronics (ICCE), IEEE, 2021, pp. 1–6.

[10] Chao Y., Xiaojia X., Wang C., Towards real-time path planning through deep reinforcement learning for a UAV in dynamic environments, Journal of Intelligent & Robotic Systems 98 (2020) 297–309.

[11] W. Zaher, A.W. Youssef, L.A. Shihata, E. Azab, M. Mashaly, Omnidirectional-wheel conveyor path planning and sorting using reinforcement learning algorithms, IEEE Access 10 (2022) 27945–27959.

[12] Lidong Zhang, Zhixiang Liu, Youmin Zhang, Ai Jianliang, Intelligent path planning and following for UAVs in forest surveillance and fire fighting missions, IEEE, 2018, pp. 1–6.

[13] D.M. Ranaweera, K.T.M. Udayanga Hemapala, A.G. Buddhika, P. Jayasekara, A shortest path planning algorithm for PSO base firefighting robots, in: 2018 Fourth International Conference on Advances in Electrical, Electronics, Information, Communication and Bio-Informatics (AEEICB), IEEE, 2018, pp. 1–5.

[14] Runnan H., Chengxuan Q., Jian Ling L., Lan X., Path planning of mobile robot in unknown dynamic continuous environment using reward-modified deep Q-network, Optimal Control Applications and Methods 44 (3) (2023) 1570–1587.

[15] V. Nguyen, T. Manh, C. Manh, P. Tiến, M. Van, D. Ha Thi Kim Duyen, D. Nguyen Duc, Autonomous navigation for omnidirectional robot based on deep reinforcement

learning, International Journal of Mechanical Engineering and Robotics Research (2020) 1134–1139, https://doi.org/10.18178/ijmerr.9.8.1134-1139.
[16] P. Nancy, S. Muthurajkumar, S. Ganapathy, S.V.N. Santhosh Kumar, M. Selvi, K. Arputharaj, Intrusion detection using dynamic feature selection and fuzzy temporal decision tree classification for wireless sensor networks, IET Communications 14 (2020) 888–895.
[17] M. Selvi, K Thangaramya, S. Ganapathy, K. Kulothungan, H. Khannah Nehemiah, A. Kannan, An energy aware trust based secure routing algorithm for effective communication in wireless sensor networks, Wireless Personal Communications 105 (2019) 1475–1490.
[18] R.S. Moorthy, P. Pabitha, Accelerating analytics using improved binary particle swarm optimization for discrete feature selection, The Computer Journal 65 (10) (2022) 2547–2569.
[19] K. Nivitha, P. Parameshwaran, C-DRM: coalesced P-TOPSIS entropy technique addressing uncertainty in cloud service selection, Information Technology and Control 51 (3) (2022) 592–605.
[20] S.V.N. Santhosh Kumar, Y. Palanichamy, M. Selvi, Energy efficient secured K means based unequal fuzzy clustering algorithm for efficient reprogramming in wireless sensor networks, Wireless Networks 27 (6) (2021) 3873–3894.

About the authors

S. Muthurajkumar received an ME in Computer Science and Engineering from Anna University, Chennai, where he has completed a PhD, and he is working as an Associate Professor in the Department of Computer Technology, MIT Campus, Anna University, Chennai. He has published more than 35 articles in journals and conferences. His areas of interest are Cloud Network Security, Cloud Computing, and Data Mining.

B. Rahul received a BE in Computer Science and Engineering from Anna University, Chennai.

L.S. Sanjay Kumar received a BE in Computer Science and Engineering from Anna University, Chennai.

E. Gokkul received a BE in Computer Science and Engineering from Anna University, Chennai.

CHAPTER ELEVEN

Delineating computational intelligence during epidemic emergencies and outbreaks

Preetha Evangeline David[a], V. Vivek[b], and P. Anandhakumar[c]
[a]Department of Artificial intelligence and Machine Learning, Chennai Institute of Technology, Chennai, Tamil Nadu, India
[b]School of Computer Science & Engineering, Faculty of Engineering & Technology, JAIN (Deemed-to-be University), Bangalore, India
[c]Department of Information Technology, Madras Institute of Technology, Anna University, Chennai, Tamil Nadu, India

Contents

Abstract

As of April 18, 2021, the coronavirus or COVID-19 virus has affected 219 countries and territories, killed over 3 million people, and infected over 141 million people globally. The purpose of this literature review is to explore how augmented reality (AR) and artificial intelligence (AI) can help to stop the spread of a disease from becoming a pandemic by informing health agencies and national governments about when to implement healthy measures, close their borders, and restrict travel sooner in the future. Research shows that AR is an excellent tool for capturing incidents in three dimensions (3D) and has the ability to maneuverer data to be viewed and understood in all directions. Also, AI has the ability to analyze vast data, learn to improve itself in order to predict future situations, and make deductive or data-reliant decisions in real time. Despite much research on both AR and AI in the health field, this research focuses on how to use these tools to make quicker and better decisions than we did in 2020 for COVID-19.

Advances in Computers, Volume 132
ISSN 0065-2458
https://doi.org/10.1016/bs.adcom.2023.10.001

1. Introduction

In March 11, 2020, the World Health Organization (WHO) declared that the novel coronavirus (COVID-19) disease had become a pandemic [1] because it had affected about 123 countries; killed about 5,000 people globally; and infected about 132,000 people, mainly in Asia, Europe, the Middle East, and now the United States [2]. COVID-19 is a viral disease that started in Wuhan, China in December 2019; it is a respiratory illness that spreads between people; it has symptoms of cough, fever, as well as shortness of breath that can lead to death; and it takes about 14 days for symptoms to emerge after exposure to the virus [3]. COVID-19 became a pandemic just within 3 months and has affected millions of workers, travelers, the elderly, and loved ones in many countries [4]. COVID-19 virus in the air from a cough or sneeze can still infect people after 3h, over 3 days from plastics and stainless steel, up to 24h from cardboard, and up to 4h from copper [5]. This means that wearing a medical mask is essential, especially when you go outside for a few hours, as well as washing your hands regularly and wiping most-used surfaces with disinfectant wipes frequently.

COVID-19 can be transmitted from animals to humans, especially from cats and camels, and can also lead to kidney failure and pneumonia if hands are not washed properly and regularly, nose and mouth are not covered properly or distance is not maintained, especially when around those coughing and sneezing, or when meat or eggs are not cooked thoroughly. Among the infected in China, about 18% are in severe or critical condition; only 2.3% have died [6]. The death rate for people above 60 years of age ranges from 3.6% to 14.8%, the death rate for those with pre-existing conditions (cardiovascular, diabetes, and cancer) ranges from 5.6% to 10.5%, and about 51% of the cases are men [6]. It seems it is best to keep some social distance from animals and sick people, as well as keep the elderly and those with pre-existing health issues away from the general population.

In March 17, 2020, the African continent had the lowest cases of COVID-19 infection, with 347 infections out of over 169,000 globally (0.2%) and just seven deaths out of about 6,500 globally (0.01%) because they implemented travel bans, against WHO's advice; restricted immigrants from highly infected countries to sustain tourism; encouraged social distancing and good hygiene; closed work and schools; and encouraged border closure [7]. Putting wisdom, as well as value for both long life and good health ahead of money, travel, and trade when making policies against the spread of a disease outbreak is paramount. A sick or dying person cannot enjoy the pleasures from money, travel, and trade. As of January 8, 2021, the United States alone had passed 4,000 deaths in one day [8], and, as of April 18, 2021 has almost 32 million affected and almost 600,000 deaths [9]. There is a justified need for citizens to put pressure on their workplaces, media, as well as travel and health agencies to demand that the government close national borders as well as restrict travel to and from infected countries within a week of an epidemic spreading to another country until the disease is under control.

Many major sporting events around the world were either cancelled or postponed, which includes the Olympic Games, the National Basketball Association, the National Hockey League, Major League Baseball, Major League Soccer, NCAA's March Madness, the PGA Tour for Golf, and even European soccer's Champions League [10]. By January 2021, many of these sporting events had resumed, but coaches and players are required to wear masks on the bench and the games are played in an empty stadium or arena, which greatly reduces the revenue for the sport organizations. Due to the COVID-19 pandemic, the stock market in the United States has suffered its worst Dow Jones Industrial Index since 1987 after losing about 2350 points, which is about 10% [11]. In case of disease outbreak, global citizens should try to take the lead and be proactive in self-travel bans and self-social distancing from work as well as sporting events, instead of reactively relying on the government and health organizations.

Local officials in Florida, USA proactively moved to shut down their beaches to prevent large crowds and spring breakers from gathering during the COVID-19 outbreak because the governor refused to close the beaches officially [12]. People in one state or country in a digital age should be able to reach others in real time strategically in other states, countries, and continents about a disease outbreak that can affect the entire country or continent within a week to prevent a disease outbreak from becoming an epidemic or a pandemic. Despite the fact that there is some research on AR and AI in healthcare and diseases, there is limited study on using both AR and AI together toward stopping the spread of an epidemic becoming a pandemic. Now that we have discussed COVID-19 as a pandemic, we will explore the literature on how AR and AI can be used together to stop a disease from becoming a pandemic. Then we will present the relationship model, methodology, and discussion. Lastly, limitations and the conclusion will be addressed.

2. Literature review

2.1 Augmented reality

AR is a user interface technology that enhances a user's physical environment by augmenting it with virtually created computer data of videos, images, sounds, graphics, global positioning systems (GPS), and texts retrieved from various sensors, and it permits the immersive capability to view content, navigate, communicate information, and change the way a user interacts with the physical environment [13]. AR is a new approach that is integrated into any information or technological system to enhance information in the real world, as well as our daily lives in both an interacting and cognitive way by tracking reference points within the data, capturing and coordinating digital data, and displaying virtual information combined with the physical environment [14]. A new international or inter-continental integrated data-sharing protocol should be created to allow WHO and local health organizations to be able to use AR to capture images and sounds of people from cameras and speakers of smartphones, digital car videos, and digital speakers in any location in the world to observe their health conditions in real time. They can also use AR to communicate with local authorities to confirm or predict what is actually going on about the health issues in any part of the world in real time.

Many organizations expect the combination of big data and AR as a necessity because AR's ability to visualize data will enhance and improve development of creative solutions for needs-based concerns, organizational business strategy, mobile or smartphone technology, analysis of vast data from the internet of things (IoT), and social media in real time [14]. AR aids users to evaluate products or situations efficiently in 3D perspective before making a decision [15]. The new international or inter-continental integrated data- sharing protocol should encourage digital and social media users to allow WHO and local health organizations to access their digital data to view health concerns in any part of the world. The digital data can be viewed in AR's 3D from various IoT connected to the internet in order to come up with a strategy or solution to prevent an epidemic or a pandemic in real time. They can also retrieve and analyze vast data from social media conversations in order to make a health decisions for that locality or nation in real time.

AR user's insight of information quality (accuracy, objectivity, reliability, and relevance of content) and visual quality (appealing, clear, and concrete presentation of 3D graphics) directly influences both perceived diagnosticity (usefulness to learn and understand the situation with functional feature use) and satisfaction (psychological and emotional opinion that exceeds expectation after usage or experience); perceived diagnosticity directly influences satisfaction; and satisfaction directly influences loyalty, which is the user's repeated usage (behavioral) or both recommendation and commitment to brand (attitudinal) [15], as shown in Fig. 1. WHO and local health organizations can get accurate and relevant content from AR that has clear and appealing 3D graphics, which can be turned into useful health information that will lead to a satisfactory decision or solution.

2.2 Artificial intelligence

AI is an intelligent system of tools, techniques, and algorithms that has the ability to think, learn, as well as decide and augment work, which requires the processing of machine learning ability (algorithms for learning), natural language (analyzing human language), and machine vision capability (algorithms for image analysis) [16]. AI learns from past experiences and data in order to develop intelligent solutions, has the ability to learn and improve itself for knowledge-based tasks, and is excellent for analytical decision making, while humans are necessary for intuitive decision making [16]. WHO and local health organizations can use AI to learn about past epidemics and pandemics, how they were resolved and the mistakes that were made, as well as predict what needs to the done differently in future to terminate diseases before they become an epidemic or pandemic. They can also use AI to predict the next city to be affected, analyze what strategies are needed to prevent the disease from spreading to those locations, and develop an intelligent solution or recommendation to prevent both an epidemic and a pandemic.

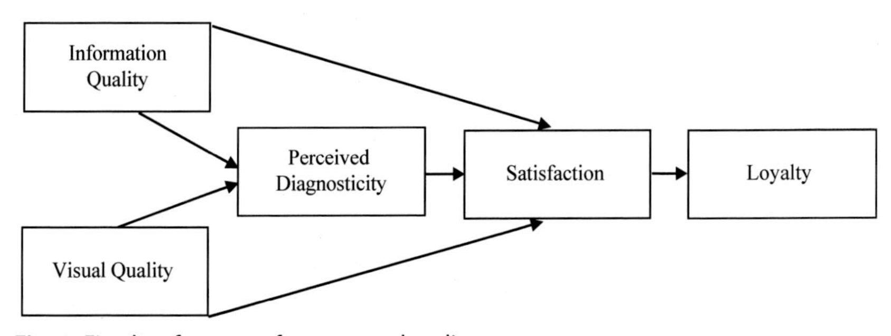

Fig. 1 Five key features of augmented reality.

The analytical approach of AI depends on the analysis of knowledge based on conscious reasoning as well as logical deliberation, but lacks the comprehension of common-sense and uncertain conditions, while the intuitive approach of humans depends on their instinct, gut feeling, and previous experiences, but has the edge of being creative and imaginative when it comes to decision making [16]. Fusing the ability of AI to analyze numerous data in real time with the edge of human intuition and discerning judgement is known as hybrid intelligence (HI) [17], as shown in Fig. 2. AI assists in upgrading human decisions by supplying predictions, while humans aid AI to learn the current machine learning models, so HI enables humans to gain from AI predictive ability and then humans utilize their intuition, imagination, and creativity to make decisions, usually based on AI's prediction, which has no discrimination of ageism, sexism, or racism (Dellermann et al., 2019). WHO and local health organizations can depend on updated or current expert systems models or HI to imitate the behaviors of experts that successfully dealt with past epidemics and pandemics, as well as apply the same methodologies without bias in their decisions and implementation of strategies in real time.

An expert system is a computerized HI and a form of AI that copies human expert behavior by retrieving and utilizing human experts as both data and rules within a computer program, which can be used to solve very complex conditions [18,19]. AI provides data-reliant reasoning from past cases; rule-based reasoning in an attempt to provide a new theory and deductive judgement; and biometrics for identification purposes, according to the physiological or behavioral characteristics for verifying and authenticating identities [19,20]. WHO and local health organizations can depend on AI, expert system models, and HI to make intelligent decisions in real

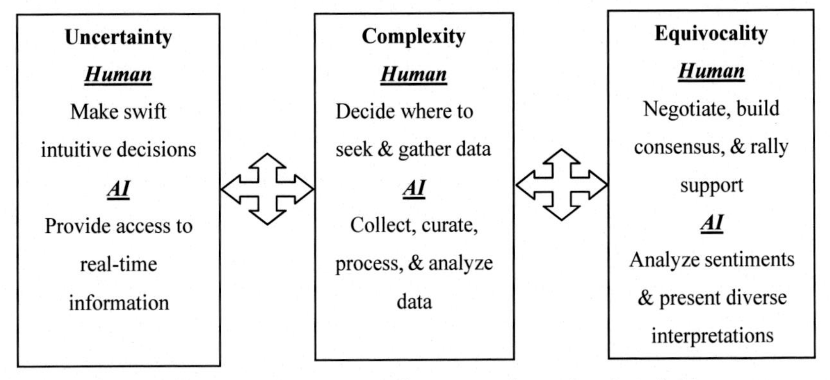

Fig. 2 Hybrid intelligence (Human & AI) decision making situations [16].

time, based on past disease outbreaks, rules for deductive reasoning, and biometrics to verify and authenticate disease and its characteristics of spreading.

In the health industry, AI can retrieve data from public health surveillance, monitor disease outbreaks in real time, forecast and get briefings from government as well as health organizations [21]. AI can provide accurate predictions and readings in the future with effective data sharing [22], but AI is currently not efficient in epidemiology, pharmaceutical, and diagnostic perspectives because of lack of data or too much unintegrated data that needs a balance between data privacy and human-AI interaction in public health [23]. Government and health organizations should be able to use the new international or inter- continental integrated data-sharing protocols from all information and communication technologies, such as smartphones and wearables, and various platforms such as social media, in order to integrate vast data in real time for adequate preventive AI reading, prediction, and forecasting.

AI can identify ongoing outbreaks within weeks, identify existing medication or discover new therapeutic options within months, and may take decades to have standardized protocols for sharing data and information during a health crisis [21]. AI can diagnose and prevent further disease outbreaks through containment as well as mitigation in order to sustain economies [24]. Through adequate data sharing, disease outbreaks can be identified in time and provide both accurate molecular cure, as well as preventive lifestyle strategies sooner if proper data-sharing protocols between all networks, platforms, and communication technologies are in place now with proper privacy and preventive controls.

We need to cover the gaps in current health policies by designing and implementing disaster and emergency policies that are highly integrated, use risk assessments to forecast future events, and plan in advance to resist and respond to any disaster in real time [21]. For efficient data sharing, there must be effective data protocols for sharing data across various networks and systems, which guarantees preventive and privacy controls, especially for medical data [22]. Preventive and privacy controls are essential for both AI and AR to become more acceptable to both governmental and health agencies, as well as to global citizens.

3. Method

Since COVID-19 is a new disease, this study relied on medical experts in reliable news sources to provide updates in real time since

March 2020, and did not want to make the content too medical or scientific. A total of 11 news sources were used to understand the nature of the disease, the health measures required to survive it, and the flaws of both national governments and health organizations to make a decisive decision sooner. A total of three articles were used to explain how AR works and its uses, and nine articles were used to understand the purpose of AI and its usefulness. The two medical sources from WHO and John Hopkins were used to explore what COVID-19 is and how to keep individuals safe from it.

The goal of this paper is to explore through a literature review of both experts in news outlets and academic journals how AR and AI can help health organizations stop the spread of an epidemic from becoming a pandemic, with the collaboration of national government, travel companies, digital companies, and even employers. Someone has to have the authority to stop business as usual when it comes to an epidemic disease becoming a pandemic, because a pandemic seems to kill and affect people faster than war, so an epidemic becoming a pandemic should be seen as a prioritized event to be controlled within a few months, rather than the pandemic controlling the entire world and our daily lives for years without end in the near future.

The strategy is to have WHO and local health agencies collect vast digital data from hospitals, health professionals, families of the sick or dead, and individuals in the epidemic region as soon as possible and repeatedly until AI can analyze the vast digital data and provide a credible and reliable decision that WHO, national governments, travel agencies, employers, and individuals should follow. This paper presents the view that the healthy person is the wealthiest and wisest person because the dead or the sick without hope of survival against a new pandemic has no use of the economy, travel, or trade.

3.1 Relationship model

Due to the issues of privacy with AR and bias with historical data in AI, it is recommended that smartphone services and social media companies require that customers agree to share their digital data of disease outbreak or epidemics in any region of the world with WHO and local health organizations around the world to prevent a pandemic crisis. This request should be based on the new international or inter-continental integrated data- sharing protocol to combat diseases, emergencies, and disasters in real time or in weeks rather than in months or years. The spread of COVID-19 in 2020 was a global failure of WHO, national governments, and health organizations.

The model in this study recommends a global or inter-continental data-sharing protocol that allows WHO and local health agencies to retrieve any digital epidemic video, image, graphics, or sound from the location of an epidemic in real time for instant AI analysis. AI can rely on deductive reasoning from past pandemics rather than relying on the past failed responses, and can also use biometrics to identify possible remedies based on detected symptoms to make vaccine creation faster. Regardless of the consequences to the economy, travel, entertainment, or trade, once AI determines that the epidemic could become a pandemic and provides safety measures to follow, WHO and local health agencies should have the power to recommend the closure of borders and stop travel immediately until the epidemic is under control. Local health agencies are to ensure that national governments are involved in the pandemic prevention exercise.

The model in Fig. 3 is founded on the integrated models in Fig. 1 (Features of AR) and Fig. 2 (Benefits of HI), which are to integrate both AR and AI to achieve the prevention of pandemic outbreaks in the future. Integrating AR with AI is simply using AI to analyze digital data provided by AR. There are a total of eight stages. Stages 1-3 display the role of AR, stages 4–6 show the role of AI, and stages 7–8 present both satisfaction and loyalty, which are derived from the first six stages.

The first three stages show how WHO and local health organizations may use AR. The second three stages present how AI may be used by them, along with AR. Stages 7–8 show how satisfied and loyal both government and citizens are with the use of both AR and AI to prevent a disease from becoming a pandemic, regardless of the types of people in the affected location.

The use of AR and AI, as shown in Fig. 3 below, to prevent diseases from becoming a pandemic is, in fact, a cycle that moves from AR to AI to both satisfaction and loyalty, then returns to AR. This shows that the success of preventing diseases from becoming a pandemic through the use of both AR and AI together should be patronized by national governments, health organizations, and employers, if it leads to an accurate and successful prevention of epidemic and pandemic, regardless of what part of the world is affected or the types of people that reside there.

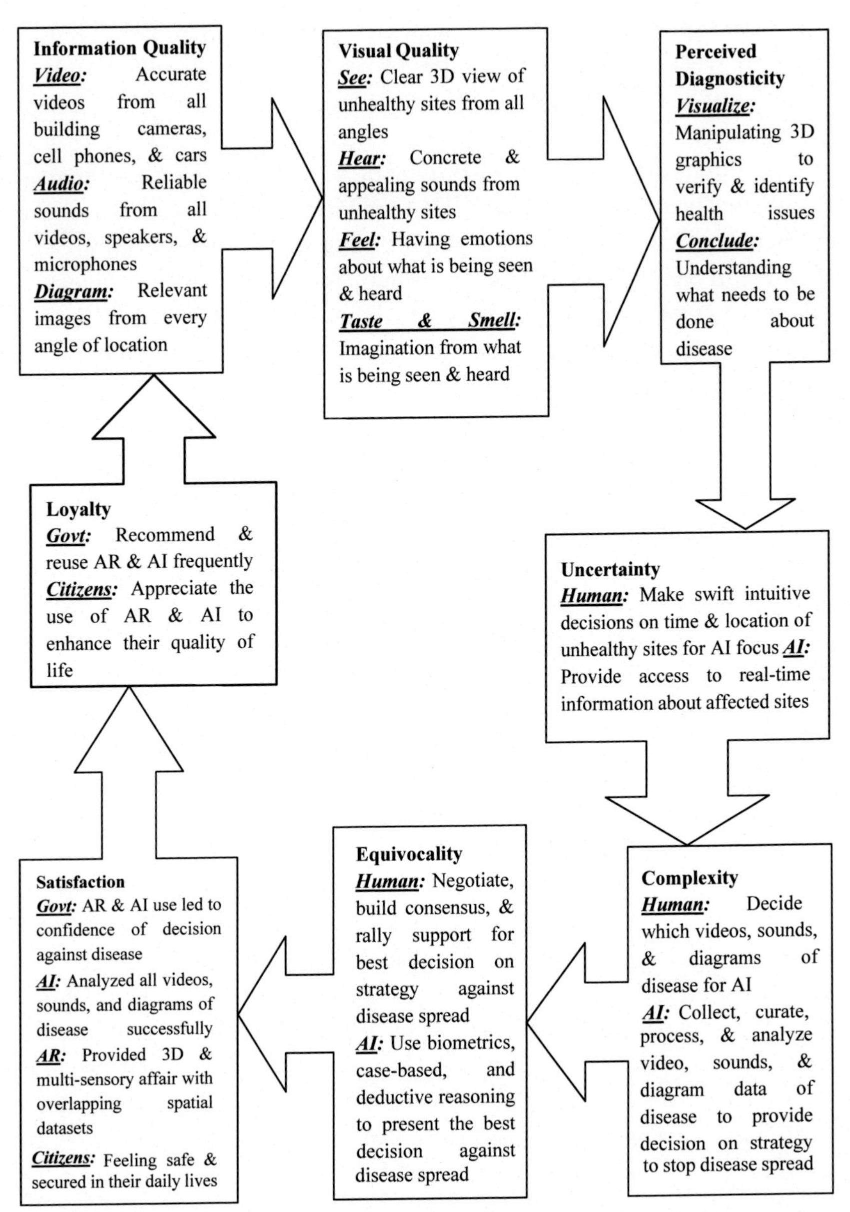

Fig. 3 Combining augmented reality and artificial intelligence cycle to prevent pandemic outbreak.

4. Results

It is assumed that in the future, an international and inter-continental integrated data-sharing protocol will be implemented to give WHO and local health agencies the right to recommend closure of borders, travel bans, and healthy measures against any epidemic from becoming a pandemic as soon as they have AR and AI evidence through integrated networks, systems, and platforms to show that it is similar or worse than the COVID-19 pandemic. National government, travel companies, employers, and individuals are to collaborate with the law or face harsh penalties for endangering humanity. The integrated AR and AI cycle for preventing a pandemic outbreak, as presented in Fig. 3.

Stage 1 is for Information Quality: WHO and local health organizations can use AR to acquire reliable audio, accurate videos, and relevant pictures from car and smartphone cameras, street and building cameras, sounds from digital speakers and microphones (such as Apple's Siri, Amazon's Alexa, and Google Assistant), as well as images about the location of disease from all angles and directions.

Stage 2 is for Visual Quality: WHO and local health organizations can use AR to be there at the scene despite the subject being absent or far away in the real world, but there is still a need to validate what was seen and heard in order to prevent any bias. At this stage, the multisensory and multi-dimensional characteristics of AR are extremely important.

Stage 3 is for Perceived Diagnosticity: WHO and local health organizations are absorbed in the combined benefits of both information quality and visual quality, which gives them the advantage of playing with the functional features of AR to manipulate videos, sounds, images, and graphics in order to match the AR experience with what is actually happening at the scene of the health crisis.

Stage 4 is for Uncertainty: This is where WHO and local health organizations become AI users by integrating their AR experience with AI in order to attain a factual conclusion and make a credible decision without bias. This is the genesis of HI, because WHO and local health organizations become AI users that use their human intuition, imagination, and creativity to inform AI about the time, location, directions, angles, and coordinates of disease affected areas to focus on in order for AI to furnish real-time analysis of the disease and affected locality.

Stage 5 is for Complexity: WHO and local health organizations decide the videos, graphics, sounds, and images to be analyzed for adequate decision making, and AI uses its analytical characteristics to classify the vast data in order to finalize the decision of what is actually occurring in the affected area of the disease.

Stage 6 is for Equivocality: AI uses biometrics for verification and authentication of objects. It will be a great tool to verify and identify various disease patterns and how they spread to different places in real time. AI also has case-based or data-reliant reasoning for finding similar diseases from the past and recommending the adequate preventive practices used to prevent the spread of that type of disease in the present or future. AI can also use abductive or rules-based reasoning to apply logic from expert systems or computerized HI that mimics the decision or behavioral pattern of credible health experts who successfully prevented epidemic and pandemic diseases in the past.

Stage 7 is for Satisfaction: WHO and local health organizations are expected to be satisfied that the AR has supplied all the relevant digital videos, sounds, and graphics required to make a verifiable decision about the disease and the affected locality. WHO and local health organizations are also satisfied and assured that they supplied AI with adequate models for the actual location and time of the disease outbreak to be analyzed, as well as reliable sounds, relevant pictures, and accurate videos for objective decision making. Citizens are expected to be happy that both AR and AI are being used to keep them safe and secure from the spread of the disease and can enjoy their daily lives.

Stage 8 is for Loyalty: National governments, health organizations, and employers may believe that AR and AI together is a dependable way to factually find out what type of disease outbreak is affecting an area and recommend preventive strategies to stop the spread of that disease from becoming a pandemic. National governments, health organizations, and employers may be so pleased with both AR and AI that they would recommend them to other organizations and reuse them as often as needed. Also, citizens may feel both safe and secure in their localities because their governments, health organizations, and employers are doing everything possible with the aid of both AR and AI to guarantee that their lifestyles are not interrupted by the spread of the disease.

5. Limitations

This study relied on research done by medical experts in news outlets (about 40%) and health organizations about COVID-19, as well as academic journals about AR and AI because COVID-19 is a new disease and the author did not want to make the content too medical or scientific to understand. The study could have been a qualitative research method with interviews of experts, but a literature review was used to show the possibility and capability of using both AR and AI to prevent the spread of a disease from becoming a pandemic.

Triangulation was not done to investigate the privacy laws of how AR could use private data from smartphones, car cameras, and building cameras to monitor diseases in real time. Also, international laws were not studied to understand how health organizations can be permitted to monitor or get past data from local databases within another country or territory. More research is needed to see if AI can provide adequate recommendations for WHO and local health organizations based on data from AR that are acceptable for strategic implementation and achieve the desired result of preventing a pandemic.

6. Conclusion

In summary, for both AR and AI to be effective and efficient tools in combating disasters, emergencies, and diseases from spreading into a pandemic, there must be a new and accepted international and inter-continental integrated data-sharing protocol to allow the sharing of integrated data across and between networks, systems, and platforms globally. National governments, health organizations, and first responders should be able to receive integrated data in the form of video, sound, text, emails, GPS, images, and graphics in real time while maintaining privacy controls to make adequate decisions.

This study shows that AR is capable of retrieving data in real time and can be manipulated in 3D to give WHO and local health organizations a good understanding of the nature of disease in various locations around the world in real time. AR can provide very clear, relevant, and concrete digital data in terms of videos, audios, graphics and GPS coordinates, where WHO and local health organizations can be involved to articulate the reality of the disease situation. AR provides information quality, visual quality, diagnosticity, satisfaction, and loyalty characteristics for their users.

Research shows that AI is excellent for gathering vast amounts of data and analyzing them in real time. AI also has the ability to learn to improve its predictions for the future and provide updated decisions or recommendations for national governments, health organizations, and employers about preventing the spread of disease, as well as protecting their citizens from a pandemic. AI can work with humans to form HI, which can be used to make decisions from past disease experience, and use rules and theories to deduct reasoning for new and future diseases, and biometrics to verify and authenticate disease types.

References

[1] B. Chappell, Coronavirus: COVID-19 Is Now Officially A Pandemic, WHO Says, 2020. http://www.npr.org. D. Dellermann, P. Ebel, M. Söllner, J.M. Leimeister, Hybrid intelligence, Bus. Inf. Syst. Eng. 61 (5) (2019) 637–643.
[2] W. Feuer, N. Higgins-Dunn, B. Lovelace, Europe is Now The 'Epicenter' of The Coronavirus Pandemic, WHO Says, 2020. http://www.cnbc.com.
[3] K. Sauer, T. Harris, An effective COVID 19 vaccine needs to engage T cells, Opin. Article 11 (2020) 1–6.
[4] H. Pattersson, B. Manley, S. Hernandez, Tracking Covid-19's Global Spread, 2021. http://www.cnn.com. L.M. Sauer, What is Coronavirus? 2020. http://www.hopkinsmedicine.org.
[5] G. Emery, Coronavirus Can Persist in Air for Hours and On Surfaces for Days: Study, 2020. http://www.reuters.com.
[6] L.T. Vo, Coronavirus Death Rate by Age, 2020. http://www.buzzfeednews.com. World Health Organization (WHO), Coronavirus, 2020. http://www.who.org.
[7] B. Adebayo, African Countries Shut Doors Against Europe, America to Combat Coronavirus, 2020. http://www.cnn.com.
[8] M. Holcombe, D. Andone, The US Reported More Than 4,000 Covid-19 Deaths in One Day For the First Time Ever, 2021. http://www.cnn.com.
[9] S. Hernandez, B. Manley, S. O'Key, H. Pattersson, Tracking Covid-19 Cases in The US, 2021. http://www.cnn.com.
[10] K. Badenhausen, Coronavirus Update: Sporting Events Canceled Or Postponed By The Pandemic, 2020. http://www.forbes.com.
[11] E. Torres, C. Thorbecke, Stock Market Surges Day After Worst Lost Since 1987, 2020. Retrieved from http://www.nbcnews.go.com.
[12] E. Fieldstadt, Coronavirus Comes For Spring Break, 2020. http://www.nbcnews.com.
[13] K.C. Brata, D. Liang, An effective approach to develop location-based augmented reality information support, Int. J. Electr. Comput. Engin. 9 (4) (2019) 3060.
[14] D. Aslan, B.B. Çetin, İ.G. Özbilgin, An innovative technology: Augmented reality based information systems, Procedia Comput. Sci. 158 (2019) 407–414.
[15] J. Yoo, The effects of perceived quality of augmented reality in mobile commerce: An application of the information systems success model, Informatics 7 (2) (2020) 14.
[16] M.H. Jarrahi, Artificial intelligence and the future of work: Human-AI symbiosis in organizational decision making, Bus. Horiz. 61 (4) (2018) 577–586.
[17] D. Dellermann, P. Ebel, M. Söllner, J.M. Leimeister, Hybrid intelligence, Bus. Inf. Syst. Eng. 61 (2019) 637–643.
[18] B. Abu-Nasser, Medical expert systems survey, Int. J. Eng. Inf. Syst. 1 (7) (2017) 218–224.
[19] R.W. Campbell, Artificial intelligence in the courtroom: The delivery of justice in the age of machine learning, Colo. Tech. LJ 18 (2020) 323.
[20] E. Nissan, Digital technologies and artificial intelligence's present and foreseeable impact on lawyering, judging, policing and law enforcement, AI Soc. 32 (3) (2017) 441–464.
[21] N.L. Bragazzi, H. Dai, G. Damiani, M. Behzadifar, M. Martini, J. Wu, How big data and artificial intelligence can help better manage the COVID-19 pandemic, Int. J. Environ. Res. Public Health 17 (9) (2020) 3176–3183.

[22] Z. Allam, G. Dey, D.S. Jones, Artificial intelligence (AI) provided early detection of the coronavirus (COVID-19) in China and will influence future urban health policy internationally, AI 1 (2) (2020) 156–165.
[23] W. Naudé, Artificial Intelligence Against COVID-19: An Early Review, 2020, pp. 1–12.
[24] M.L. Jibril, U.S. Sharif, Power of Artificial Intelligence to Diagnose and prevent Further Covid-19 Outbreak: A Short Communication, 2020, pp. 1–8.

About the authors

Dr. Preetha Evangeline David is currently working as an Associate Professor and Head of the Department in the Department of Artificial Intelligence and Machine Learning at Chennai Institute of Technology, Chennai, India. She holds a PhD from Anna University, Chennai in the area of Cloud Computing. She has published many research papers and Patents focusing on Artificial Intelligence, Digital Twin Technology, High Performance Computing, Computational Intelligence and Data Structures. She is currently working on Multi-disciplinary areas in collaboration with other technologies to solve socially relevant challenges and provide solutions to human problems.

Dr V. Vivek is working as an Associate Professor in the Department of Computer Science and Engineering at Jain University, Bangalore. He also serves as a Deputy Director (Student and Industry Relation). He has 14+ years of Teaching and Research experience in the field of Cloud Computing and Networking. He holds a doctoral degree from Karunya Institute of Technology and Sciences. He has published many Research articles in high reputed journals and has mentored sponsored funded projects.

Dr. P. Anandhakumar is a professor in the Department of Information Technology at Anna University, Chennai. He has completed his doctorate in the year 2006 from Anna University. He has produced 17 PhD's in the field of Image Processing, Cloud Computing, Multimedia technology and Machine Learning. His ongoing research lies in the field of Digital Twin Technology, Machine Learning and Artificial Intelligence. He has published more than 150 papers indexed in SCI, SCOPUS, WOS, etc.

CHAPTER TWELVE

ClubNet: Deep learning model for computation, calibration and estimation of biotic stress in crops

Preetha Evangeline David[a] and P. Anandhakumar[b]

[a]Department of Artificial Intelligence and Machine Learning, Chennai Institute of Technology, Chennai, Tamil Nadu, India

[b]Department of Information Technology, Madras Institute of Technology, Anna University, Chennai, Tamil Nadu, India

Contents

Abstract

Biotic stress is purely defined as the devastation of crops through another living organism. The concept of Agricultural viability is closely related to the decisive restraint of pests and pathogens. Viable agriculture leads to the growth in new technologies followed by increased productivity, reduced environmental influences and receptiveness to farmers. Deep learning methodologies and computer visualization results in accurate identification of stress causing factor. Right solutions can be initiated to intensify the problem. The contribution of this research lies in diagnosing and estimating the level of severity caused by biotic factors on crops. Bacteria and other biological factors infects leaves, shoots and berries. The symptoms appear as minute water soaked spots on the lower surface of the leaves, especially along the main and lateral veins.

Advances in Computers, Volume 132
ISSN 0065-2458
https://doi.org/10.1016/bs.adcom.2023.08.004

Mostly these spots coalesce and form larger patches. Severely infected leaves give a blighted appearance which leads to cane immaturity, water berries, cluster tip wilting, shot berries, uneven ripening and post-harvest berry drop This paper proposes a ClubNet based on Convolution Neural Network which is a multi-tasking system and additionally delve into data augmentation techniques to have a booming potential system. The proposed ClubNet facilitates extorting complimentary discriminative features. Proposed Model resulted in an accuracy of 96.04% for biotic stress and 93.22% of severity estimation of stress levels. The proposed ClubNet has been compared with GoogleNet, VGGNet, DenseNet and ResNet which proved the proposed Model resulted in average accuracy of 98.19, an average precision of 96.41% which was the highest compared to other models. The proposed approach is highly suitable for modern agricultural practices.

1. Introduction

Grapes are known to be one of the economically cultivated fruit crop throughout the world which is widely used for wine, non-fermented drinks and fresh fruit. However, grapes are susceptible to various biotic stress such as Black_rot, Esca or the Black_Measles, Leaf_Blight or Isariopsis. India is among the first ten countries in the world in the production of grape. The major producers of grape are Italy, France, Spain, USA, Turkey, China and Argentina. This crop occupies fifth position amongst fruit crops in India with a production of 1.21 million tonnes (around 2% of world's production of 57.40 million tonnes). Quality of production, minimizing the loss efficiently can be achieved through possible early detection of the stress. The process of identifying and diagnosis is impartial, prone to errors and time consuming. There are certain possibilities where new diseases can cause damage to the crop and there is very much less possibility to combat them.

Technological developments and various instruments have led to automatic identification of anomalies in crops. Traditional Machine learning methodologies have greatly supported in diagnosis but lag behind due to certain limitations of their pipelined course of action (feature extraction, pattern recognition, clustering, SVM, KNN and neural networks). In contrast, the deep learning techniques tend to identify optimal pathological features automatically and also omits manual design and pipelined procedures of classifiers and feature extraction. Deep learning techniques not only support disease identification in crops but it excels in all others fields such as pedestrian crossing, biomedical image processing, face recognitions techniques, etc. because of its generalization potentiality. Coming to the field of agriculture, deep learning methods have been used widely in seed selection, identifying weeds, pest detection, summing up of fruits, future prediction based on climatic conditions, crop yield and land cover.

Convolution Neural Networks is one of the most widely used methods which has high success rate in image classification. With this idea a Club CNN has been proposed for automatic detection of biotic stress in Grape leaves.

The paper is as organized as follows, Section 2 discusses the related works, Section 3 describes the dataset and how it is pre-processed, Section 4 discourses the proposed Club CNN Model, its architecture and analyses the detection of disease using mathematical model, Section 5 deals with the experimental procedures and the obtained results, comparison graphs have been represented to significantly prove the efficiency of the proposed model, limitations and future work has been discussed, Section 6 concludes the paper.

2. Related works

As discussed in Introduction, classical Machine learning techniques were used in plant disease identification to identify the infected crops. Li et al. [1] used K- means clustering and SVM classifier for identifying mildew disease in grape plant based on 31 selective feature. Disease Identification is the primary factor to control pests and spreading of disease but estimating the severity level and quantification of the stress are two important factors which need more concentration. Bock et al. [2] discusses on the severity level on the plant tissues to predict yield and control measures. Ability to identify and relate patterns and visual evaluation is quiet challenging. Barbedo et al. [3] discussed on the important of evaluating the symptoms which is subjected in identifying the similarity of the disease and variations in the characteristics influenced by the external factors. Also this paper discusses on six important challenges that exist and classifies these problems into extrinsic and intrinsic factors. Some of the extrinsic factors include, complexity in background, variations in reflection and blurring, poorly defined boundary between heathy and infected tissues, disorders of biotic and abiotic symptoms on the same leaf, etc. Athanikar et al. [4] in his work seeked Neural Network to classify images of potato leaves into healthy and diseased. Their work showed that using BPNN could result in effective detection of diseased spots with 92% of accuracy. LeCun et al. [5] discuses on factors concerning machine learning applications which led to the significant growth in Artificial Intelligence. This led to the emergence of different models called Deep learning. Fuentes et al. [6] stated that CNN has proved state of art performance among any other methods in Deep learning. CNN has an ability to learn features from training dataset whereas other methods depends on the handcrafted features. And also the segmentation step is inbuilt in the CNN filters which furthers simplifies the process. Zhu et al. [7] analyzed on CNN in deep learning which resulted in identifying high variance in inter and intra class similarities of plant pathology. Zhang et al. [8] in his paper used GoogleNet to identify powdery mildew disease in cherry leaves and suggested that transfer learning can improve the efficiency of deep learning in plant disease identification. Ghoshal et al. [9] proposed a CNN framework which efficiently classified a set of eight stresses both biotic and abiotic. The paper considered leaves of Soyabean to identify and extract the significant features that maps exactly with the visual symptoms of the stress severity. Barbedo et al. [10] in his work used only lesions for examination instead of full leaves. Multiple lesions were identified in the same leaf but however segmenting the lesions manually was a laborious process which increased its time complexity. Manso et al. [11] proposed a threshold based method to segment the symptoms in coffee leaves. Automatic learning of features is a limitation whereas only manual features were given as input. The proposed segmentation method although showed good results, did not consider the propositional pixels which could be easily impacted by illumination and reflection factors. Mohanty et al. [12] improved deep learning models which were pre trained on ImageNet. It identified 14 crop variety and 26 diseases.

Although traditional CNN models such as AlexNet, GoogleNet, DenseNet, VGGNet and ResNet were considered for effective classification of crop diseases and also resulted in high accuracy level, there is a fact where still achieving precision is a challenge. Usage of a single model is insufficient to meet the need in terms of precision hence this paper proposed a model called ClubNet which integrates the architecture of ResNet and GoogleNet which makes it more powerful and the concept of transfer learning is inculcated which boosts up the accuracy and minimizing the training time.

3. Dataset and pre-processing

The dataset collected for this work contains leaves of grapes affected by biotic stress. Table 1 shows the plant disease dataset has been taken from an open repository of kaggle where the training data are catogorized as Black_Rot (1888 images), Esca or Black_Measles (1920 images), Leaf_Blight (1722 images), healthy leaves (1692 images) and Valid dataset consisted of Black_Rot (472 images), Esca or Black_Measles (480 images), Leaf_Blight (430 images), healthy leaves (423 images). Leaves were chosen

Table 1 Description of the dataset.

Label	Classification of disease	Training samples	Validation samples	Symptoms
1.	Black rot	1888	472	Small black lesions
2.	Black Measles	1920	480	Reddish brown strips
3.	Leaf Blight	1722	430	Brownish spots all over the leaves
4.	Healthy	1692	423	–

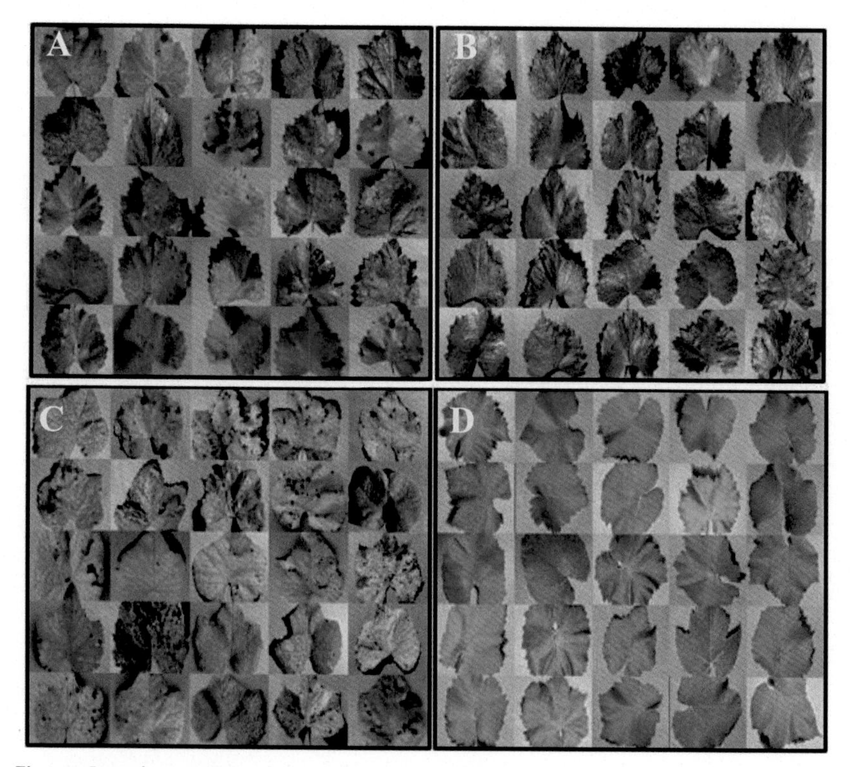

Fig. 1 Sample images of Grape leaves (A) Black_Rot, (B) Esca or Black_Measles, (C) Leaf_Blight, (D) Healthy.

to evaluate using the proposed work since the flower fruits are seasonal whereas leaves are available throughout the year and stem can hardly present the symptoms clearly and leaves are prone to sensitivity where texture, shape and color could be easily distinguished from healthy ones. The dataset are of RGB image and arbitrary in size. The leaves showed more than one type of stress with varying severity (Fig. 1).

Comparing the number of healthy samples with the infected ones, the proportion has been set to 1:2, while the class weight ratio was set to 2:1, perhaps the loss function can be negotiated during the training process. The biotic stress and its associated severity were predominantly labelled over the leaves in the dataset. A total of 2077 images showed stress of more than one type and 630 images showed stress severity of similar type. The severity level of the stress was estimated using the threshold based method and the symptoms with very poor severity level was discarded and the remaining images were ranked and ranges were fixed for healthy (<0.1%), poor (0.2–5%), below average (5.1–10%), average (10.1–15%), above average (15.1–20%) and high (<20.1%)

3.1 Elimination of over-fitting

Over-fitting is one of the obstacle faced when training a small sized dataset using a complex model. In such cases Data Augmentation methodologies could be used to eliminate over-fitting. To maximize the generalization ability of the proposed model, Mixed Augmentation technique has been applied to the original dataset to obtain fabricated set of samples combining two random images which results in generating a new image. In the method of Mixed augmentation, "a" represents set of images and "b" represents set of labels and using beta distribution of probability, the randomly value is generated. Therefore a new image is generated by the beta probability distribution as,

$$\{a, b\} = \lambda\{a_i, b_i\} + (1 - \lambda)\{a_j, b_j\}$$

where

$$\lambda \in [0, 1] \tag{1}$$

3.2 Mathematical formulation of the problem

Computing and calibrating the biotic stress in the sample leaf dataset can be formulated as a Multi-class problem. A mapping function is derived which maps the images to its corresponding labels as $f: A \rightarrow B$. The training dataset is denoted as (A_1, B_1), (A_2, B_2), $\cdots$, (A_n, B_n). To minimize the loss function the training dataset is fed as an input to the CNN where, the weights of "X" are updated continuously. The average Loss function is given calculated as, $i = 1$

$$\mathbb{K}(\omega) = -\frac{1}{4}\sum\nolimits_{i=1}^{n} B_i \ log\left(h_\omega(Ai)\right) \quad (2)$$

where A is the training sample and n is the number of input data, Bi is the label and h_ω is the prognostic label of the CNN, given the weighs ω. Here the stochastic gradient method have been used for optimizing the entire process. The gradient of the loss function is calculated as,

$$\omega t + 1 = \mu_{\omega_t} - r\nabla\mathbb{K}(\omega t) \quad (3)$$

where,

μ – Momentum of weights ωt

r – Rate of learning

Here the number of iteration are limited to that of ω, since the convolution layers are pre trained with fixed network weights. After few continuous epochs, $g : A \rightarrow B$ as a result of final hypothesis (Fig. 2).

3.3 Multi-tasking CNN

Despite of several existing CNN architectures, they all have a common goal of improving the accuracy and minimizing the complexity of the system. This research holds two objective with respect to the dataset, (i) to classify the biotic stress and (ii) calculate the severity. To satisfy these two objectives, a Multi-Tasking CNN architecture has been designed where the model is additionally added with a fully connected layer along with the existing layers. Hence the model isolates the classification blocks and shares the convolution layers (Fig. 3).

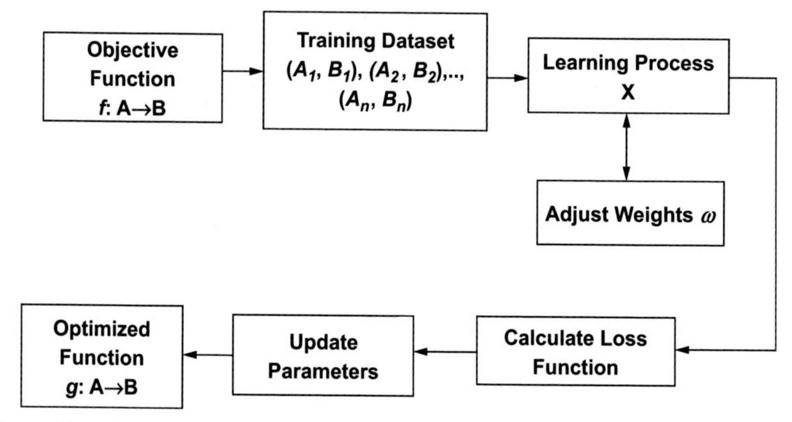

Fig. 2 Flow diagram of the problem.

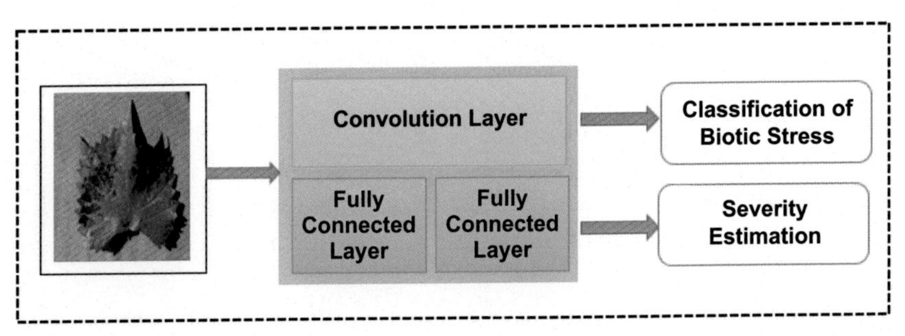

Fig. 3 Multi-tasking CNN.

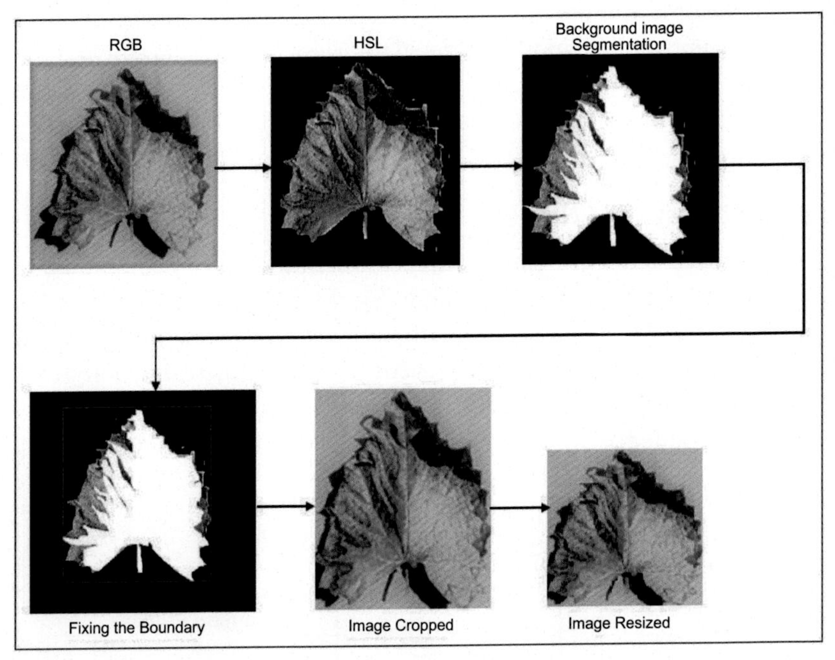

Fig. 4 Segmentation and resizing of images.

The CNN model requires an input size of the image to be $224 \times 224 \times 3$. Due to the differing size of images it is ultimately hard to detect very small sized symptoms on the surface of the leaves hence a boundary has to be fixed which eliminates the unwanted part of the leaf that does not come under infected portion. A tradition threshold based method was used to fix up the boundary using HSL values and segment the images. After segmentation process, the leaves were cropped and resized which covers most of the infected parts. For conducting the experiments the dataset was proportioned as 6:2:2 which was training set, validation set and test set (Fig. 4).

4. Proposed ClubNet model

The proposed ClubNet is based on two existing modes which are GoogleNet and ResNet 50. The significance of GoogleNet is its inception module which supports multi-scale feature extraction and combines it along with dimension reduction. ResNet on the other hand addressed the degradation challenge through its residual units. The efficiency of a CNN could be improved by increasing the parameters. GoogleNet has the potential to increase the number of units at each layer, whereas, ResNet is capable of increasing the network layer or the depth. The combination of these two models results in a renowned feature data with improvement in accuracy which would help in prediction based on combined feature sets.

The proposed model inculcated transfer learning, where the input image is resized and forwarded parallel to GoogleNet and ResNet, where the multi- Convolution filters in the Inception module and ResNet extracts the feature maps. The feature maps are then fed as input to the Global Convolution Pooling where it minimizes the chances of overfitting by reducing the number of negotiable parameters and it performs spatial and dimensionality reduction. The output from the GAP would be a two dimensional matrix followed by densely connected layer with a softmax activation function that provides the predicted objects. During classification, the neurons present in the connected layer provided connections to all learned feature maps. The combined two dimension feature matrices connected with each other matrices generating feature maps. The dropout layer are used with a rate of 0.5 and softmax classifier is used for generating the final classification result (Fig. 5).

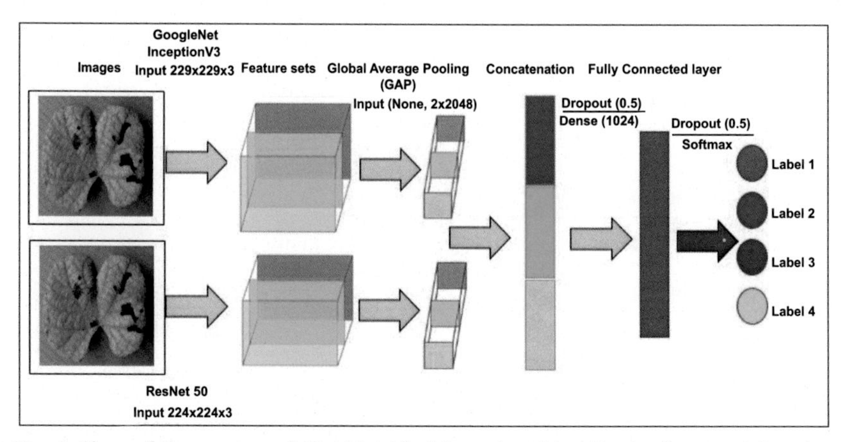

Fig. 5 Flow of the proposed ClubNet Model used to identify the best weights that minimized the loss function. SGD requires.

5. Experimental setup

The experiments were conducted on Keras v2.3.0, which is an Opensource framework that runs on top of Tensorflow. The existing models such as GoogleNet, VGGNet, ResNet, DenseNet and AlexNet were tested along with the proposed ClubNet during training phase to estimate the performance of each model. The parameters were standardized to obtain comparison results between the models. During the training phase, the corresponding weights which resulted in minimal loss function were saved for each network. Stochastic Gradient Descent (SGD) approach was training inputs of randomly fixed size which was fixed to 32 and the learning rate was set as 0.01. The test dataset were used for evaluation purpose of the saved models and results were calculated with respect to Accuracy, precision and Recall. Series of experiments were conducted to evaluate how each model responds for Biotic stress detection and severity estimation. The parameters that were used during the training are presented in Table 2.

6. Results and discussion

The experiments were performed on the Leaf dataset and results were obtained which is tabulated in Table 3. The results were classified based on Biotic stress and Severity estimation in terms of Accuracy, precision and Recall of a Multi-tasking system. The training was carried out with the training dataset instead of validation dataset to avoid over fitting. Classification was based on the True Positive (TP), False positive (FP)

Table 2 Training parameters.

Hyper-parameters	Values
No of epochs	100
Batch size	32
Optimizer	Stochastic gradient descent
Loss function	Cross entropy
Learning rate	0.01
Momentum	0.9

Table 3 Estimation of evaluation metrics for leaf dataset.

Mission	CNN models	Accuracy A1 (%)	Precision P_c (%)	Recall R_c (%)
Identification of biotic stress	GoogleNet	92.67	88.42	90.20
	AlexNet	91.56	89.14	88.78
	VGGNet	92.41	90.50	89.32
	DenseNet	93.06	91.12	89.16
	ResNet 50	95.24	92.27	91.09
	Proposed ClubNet	96.04	94.36	92.18
Estimation of severity	GoogleNet	86.24	90	88.56
	AlexNet	82.69	85.17	83.27
	VGGNet	87.53	83.12	84.92
	DenseNet	90.33	88.52	87.29
	ResNet 50	91.88	91.72	88.63
	Proposed ClubNet	93.22	90.15	89.55

and False negative (FN) values. Experimental results were obtained for all the network models and tabulated.

The Average, precision and Recall is estimated as follows:

Precision of each class "C" is calculated as,

$$Pc = TPc/TPc + FPc \tag{4}$$

Recall Rc is calculated as,

$$Rc = TPc/TPc + FNc \tag{5}$$

Average score $A1$ is calculated as,

$$A1 = \frac{2}{4}\sum\nolimits_{c-1}^{4} Rc * Pc/Rc + Pc \tag{6}$$

The result tabulated in Table 3 shows that the proposed ClubNet Model proves an accuracy of 96.04%, precision of 94.36% and Recall of 92.18% for identification of Biotic stress and for severity estimation the proposed model shows an accuracy of 93.22%, precision of 90.15% and Recall of 89.55%.

6.1 Training performed on diseased dataset

The symptom dataset consisted of images infected by the disease and did not contain any healthy leaf images. An improvised results were expected when compared to the results obtained from the leaf dataset. All the images in the diseased dataset has region of interest where the network could focus and visualize the symptoms at a greater level. Table 4 shows the evaluation of diseased dataset in terms of biotic stress.

The graph in Fig. 6 shows the average accuracy obtained for various models in terms of Training, Validation and Test images. It shows the proposed ClubNet proved better accuracy of 98.19% when compared to GoogleNet and ResNet 50 which were proved to be the existing efficient models.

The predicted results using ClubNet is depicted in Fig. 7. The results for biotic stress is found to be consistent except for the slight variance in Leaf_Blight which was due to the dataset imbalance, whereas, in severity estimation, the obtained accuracy was 93%. The proposed model could identify the healthy samples from the diseased ones without any difficulty, however mild variations occurred while classifying classes between average and high severity rates. On noticing the mild errors were found onto the main diagonal of the confusion matrix which could be negligible.

6.2 Discussion

The experimental Results of the proposed work are directly compared with five other models such as GoogleNet, AlexNet, VGGNet, DenseNet and ResNet 50 in terms of classification between biotic stress and their severity levels given the input as diseased dataset. The average Accuracy, precision

Table 4 Identification of biotic stress from a purely diseased dataset.

CNN Models	Accuracy $A1$ (%)	Precision P_c (%)	Recall R_c (%)
GoogleNet	95.61	96.13	96.27
AlexNet	93.22	92.12	94.83
VGGNet	94.74	95.66	96.77
DenseNet	95.87	93.35	95.41
ResNet 50	97.36	96.11	96.79
Proposed ClubNet	98.19	96.41	97.64

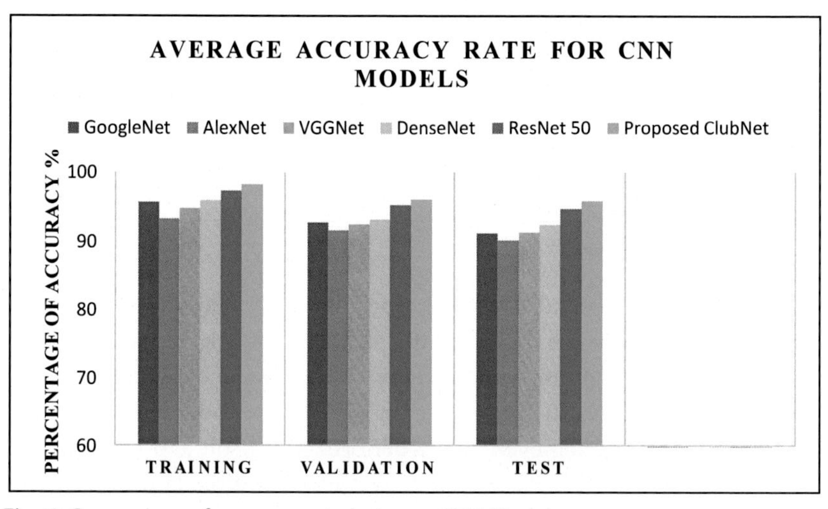

Fig. 6 Comparison of accuracy rate between CNN Models.

A

Predicted Label

acc % = 0.9604 err = 0.04

	Healthy	Black_Rot	Black Measles	Leaf_Blight
Healthy	38 1.000	0 0.000	0 0.000	0 0.000
Black_Rot	0 0.000	86 0.956	1 0.011	1 0.011
Black_Measles	0 0.000	0 0.000	147 0.987	0 0.000
Leaf_Blight	0 0.000	3 0.040	0 0.000	72 0.960

B

Predicted Label

acc % = 0.9322 err = 0.07

	Healthy	Poor	Average	High
Healthy	40 1.000	0 0.000	0 0.000	0 0.000
Poor	0 0.000	110 0.828	23 0.172	0 0.000
Average	0 0.000	5 0.098	45 0.882	1 0.020
High	0 0.000	0 0.000	6 0.375	9 0.562

Fig. 7 Confusion Matrix of the proposed ClubNet estimation results obtained using the dataset. (A) Biotic stress, (B) Severity estimation.

and the Recall value has been calculated for the overall dataset as well as pure diseased dataset. The proposed model was analyzed with the dataset of grape leaves and tested for only three diseases that are Black_Rot, Black_ Measles and Leaf_Blight. The best practice to utilize the proposed model would be testing it with random leaf samples. Statistical methods could be incorporated to estimate the general severity of each disease in various crops which would further facilitate the decision making process related to biotic stress.

The dataset used for testing purpose was associated with very less disorders and the proposed model need to be trained with different stress levels which will bring out the efficiency of the system in making inference and classification accuracy.

7. Conclusion

The work proposed a deep learning model called ClubNet which was developed from the basic of the existing models GoogleNet and ResNet 50. Grape leaf diseased dataset was used for the experimental purpose. The system is Multi-tasking which makes the model more efficient. The proposed model achieved better performance result and outperforms when compared with other CNN models in identification and classification of the disease. Accuracy, precision and Recall values were considered as evaluation metrics where the proposed ClubNet resulted in an average accuracy of 98.19%, average precision of 96.41 and recall value to be 97.64%. ClubNet resulted in 96.04% of accuracy in classifying the biotic stress and 93.22% of accuracy in identifying the severity levels of the stress. Additionally, Data Augmentation technique, elimination of over- fitting and generalization factor was induced in the system which makes it more effective.

References

[1] G. Li, Z. Ma, H. Wang, Image recognition of grape downy mildew and grape powdery mildew based on support vector machine, in: Proc international conference on computer and computing technologies in agriculture, 2011, pp. 151–162.

[2] C. Bock, G. Poole, P. Parker, T. Gottwald, Plant disease severity estimated visually, digital photography and image analysis, and by hyperspectral imaging, Crit. Rev. Plant Sci. 29 (2010) 59–107.

[3] J.G.A. Barbedo, A review on the main challenges in automatic plant disease identification based on visible range images, Biosyst. Eng. 144 (2016) 52–60.

[4] G. Athanikar, P. Badar, Potato leaf diseases detection and classification system, Int. J. Comput. Sci. Mob. Comput. 5 (2) (2016) 76–88.

[5] Y. LeCun, Y. Bengio, G. Hinton, Deep learning, Nature 521 (2015) 436–444.

[6] A. Fuentes, S. Yoon, S. Kim, D. Park, A robust deep-learning-based detector for real-time tomato plant diseases and pests recognition, Sensors 17 (2017) 2022.

[7] H. Zhu, Q. Liu, Y. Qi, X. Huang, F. Jiang, S. Zhang, Plant identification based on very deep convolutional neural networks, Multimed. Tools Appl. 77 (22) (2018) 29779–29797.

[8] K. Zhang, L. Zhang, Q. Wu, Identification of cherry leaf disease infected by Podosphaera Pannosa via convolutional neural network, Int. J. Agric. Environ. Inf. Syst. 10 (2) (2019) 98–110.

[9] S. Ghosal, D. Blystone, A.K. Singh, B. Ganapathysubramanian, A. Singh, S. Sarkar, An explainable deep machine vision framework for plant stress phenotyping, Proc. Natl. Acad. Sci. 115 (2018) 4613–4618.
[10] J.G.A. Barbedo, Plant disease identification from individual lesions and spots using deep learning, Biosyst. Eng. 180 (2019) 96–107.
[11] G.L. Manso, H. Knidel, R. Krohling, J.A. Ventura, A smartphone application to detection and classification of coffee leaf miner and coffee leaf rust, arXive (2019). prints, arXiv:1904.00742.
[12] S. Mohanty, D. Hughes, M. Salathe, Using deep learning for image-based plant disease detection, Front. Plant Sci. 7 (2016) 1419.

Further reading

[13] A. Krizhevsky, I. Sutskever, G. Hinton, ImageNet classification with deep convolutional neural networks, Adv. Neural. Inf. Process Syst. 25 (2012) 1106–1114.
[14] K. Simonyan, A. Zisserman, Very deep convolutional networks for large-scale image recognition, 2014. http://arxiv.org/abs/1409.1556.
[15] C. Szegedy, W. Liu, Y. Jia, P. Sermanet, S. Reed, D. Anguelov, et al., Going deeper with convolutions, in: Proc IEEE conference on computer vision and pattern recognition, Columbus, USA, 2014.
[16] F. Iandola, M. Moskewicz, S. Karayev, R. Girshick, T. Darrell, K. Keutzer, Densenet: implementing efficient convnet descriptor pyramids, 2014. https://arxiv.org/abs/1404.1869v1.
[17] K. He, X. Zhang, S. Ren, J. Sun, Deep residual learning for image recognition, in: Proc IEEE conference on computer vision and pattern recognition. Boston, Massachusetts, 2016, pp. 770–778.

About the authors

Dr. Preetha Evangeline David is currently working as an Associate Professor and Head of the Department in the Department of Artificial Intelligence and Machine Learning at Chennai Institute of Technology, Chennai, India. She holds a PhD from Anna University, Chennai in the area of Cloud Computing. She has published many research papers and Patents focusing on Artificial Intelligence, Digital Twin Technology, High Performance Computing, Computational Intelligence and Data Structures. She is currently working on Multidisciplinary areas in collaboration with other technologies to solve socially relevant challenges and provide solutions to human problems.

Dr. P. Anandhakumar is a professor in the Department of Information Technology at Anna University, Chennai. He has completed his doctorate in the year 2006 from Anna University. He has produced 17 PhD's in the field of Image Processing, Cloud Computing, Multimedia technology and Machine Learning. His ongoing research lies in the field of Digital Twin Technology, Machine Learning and Artificial Intelligence. He has published more than 150 papers indexed in SCI, SCOPUS, WOS, etc.

CHAPTER THIRTEEN

Automatic programming (source code generator) based on an ontological model

Preetha Evangeline David[a], S. Malathi[b], and P. Anandhakumar[c]

[a]Department of Artificial Intelligence and Machine Learning, Chennai Institute of Technology, Chennai, Tamil Nadu, India
[b]Department of Artificial Intelligence and Data Science, Panimalar Engineering College, Chennai, India
[c]Department of Information Technology, Madras Institute of Technology, Anna University, Chennai, Tamil Nadu, India

Contents

Abstract

Source AI is an AI-powered tool that can generate code in any programming language from any human language description. It can also simplify, find errors and fix them and debug your code. Automatic code generation capabilities continue to evolve within

Advances in Computers, Volume 132
ISSN 0065-2458
https://doi.org/10.1016/bs.adcom.2023.08.008

programming languages, IDEs and tools that work at compile time. This coding technique has proliferated because it can reduce mundane programming grunt work, and developers have found that it improves turnaround times and accuracy. Auto generated code usually becomes a hindrance for developers who want to tweak it later on Teams should plan to restrict these tools to only certain parts of the SDLC, such as where they can act as facilitators in smaller, less complex situations.

1. Introduction

Before organizations introduce automatic code generation, project leaders need to explicitly lay out what they mean by this development concept. Doing so will improve discussions across teams. Automatic code generation likely means one of three things to an organization:

1.1 Reflection support

Reflections in dynamic languages such as Ruby can automatically generate code within a running program. Reflection support works in a similar fashion to a smart search and automation process, which inspects the code within the language and acts in response.

1.2 Automatic code generation

Developers can use external tools required for statically typed languages—such as Go—to automatically generate source code.

1.3 IDEs

Visual Studio and similar IDEs allow developers to build logical structure declaratively using XML. This can then be translated into code. The IDE scenario falls somewhere between reflection support and external code generation. It's cleaner because of standardized code generation and the wholly managed development environment. However, this angle's potential benefits are more limited.

2. Automatic code generation use cases

Teams need to consider the trade-offs between reusable blocks of code to accomplish a task and automated code generation. Any kind of repetitive coding can slow development progress and introduce mistakes. Developers can mitigate these flaws with functions or by code generation. "These two options are actually more similar than people generally realize," Bartlett said, as they both eliminate manual coding work. At the beginning of many projects, developers might navigate a library of patterns or blueprints, select what's appropriate, fill in a few fields and the tools automatically create an application template. While this example is often associated with new applications, teams can also use it when they want to add health checks, monitoring or other capabilities. The biggest challenges relate to reducing the blueprints to only what the team needs and organizing them in an effective way.

Dependency-scanning tools automate the identification of dependencies with known vulnerabilities, which makes it easier to respond when a team discovers new security risks. In this case, the code generation automates the process of provisioning a quick fix for applications that use the dependency. Code generation can automate API development in one system to match another, Bartlett said. For example, it's often simpler to use a code generation tool to write a code handler than to do so manually, because the tool automatically analyzes the database. With automatic code generation, it's also easier to keep an API in sync with the database, since the tool can identify errors at compile time – rather than as part of a separate testing step. Automatic code generation can also simplify code migration as companies adopt newer technology stacks.

Application code needs to be kept in sync with various supporting artifacts, such as API documentation, SDK and tests. Automatic code generation tools can help automate this process to ensure changes are propagated across artifacts when updated code is pushed into production. For example, software documentation tools such as Doxygen, GhostDoc, Javadoc and Docurium can create basic documentation from annotated source code.

2.1 Extending the power of code

Many teams turn to compile time tools to automatically generate code after developers have written the main code. For example, Google has protocol buffers to serialize structured data with diverse languages and platforms. "At Google, we use protocol buffer for almost everything, from database schemas to API service endpoints". Built value tool can be used to generate the builder classes required for their preferred programming style. As a result, the code relies on well-tested libraries that are automatically updated. This reduces the manual work when changes are required.

2.2 Challenges with automation

While automated code generation removes a lot of tedious tasks from the hands of developers, there are possible perils.

Teams need to be aware that a continuous integration (CI) pipeline generates code the same way each time. Problems can occur, Sun said, when team members have local versions of code generation libraries, which results in different code generating from the same starting point. To solve this problem, standardize common automatic coding software versions and set up a CI pipeline that uses standardized code formatters, linters and lots of tests to provide a single source of truth.

Automated code generation can make programming faster in such a way that it's now harder to see potential ramifications, Bartlett said. The breaking points in programs can move drastically. "It causes minor changes to potentially have far-reaching consequences," Bartlett said. It also means a team will need to maintain, test and document the tools for metaprogramming, which involves dynamically programming tools that generate the actual code, including IDE and compile time code generation tools.

Ongoing evolution of software projects can be problematic in tools, such as Visual Studio, that make it difficult to identify generated code. Bartlett has found it useful to use a naming scheme for automatically created code, so that the generated code is easily recognized even in a file listing. However, sometimes there is crossover. Generated code sometimes needs manual modification, in which case it is no longer "generated," and requires developers to maintain it manually.

One of the most recent advancements in natural language processing (NLP) is the emergence of large language models (LLMs) that are built using vast datasets with enormous amounts of data. There are several LLMs that are available, such as Google's BERT and OpenAI's GPT-2 and GPT-3. With these models, it is possible to generate everything from simple essays to actual financial models with these models. AI startups including OpenAI, Hugging Face, Cohere, AI21 Labs are pushing the boundaries of LLM by training models with billions of parameters. Here are five AI-based code generators based on the large language models that can generate high-quality code:

2.3 OpenAI Codex

OpenAI Codex is the model based on GPT-3 that powers GitHub Copilot - a tool from GitHub to generate code within mainstream development environments including VS Code, Neovim, JetBrains, and even in the cloud with GitHub Codespaces. It claims to write code in at least a dozen languages, including JavaScript, Go, Perl, PHP, Ruby, Swift and TypeScript, and even BASH. The model is trained on billions of lines of code available in the public domain, such as GitHub repositories. OpenAI made the model available through a private beta to developers and platform companies to build tools and integration.

2.4 Tabnine

While Tabnine is not an end-to-end code generator, it puts the auto-completion feature of the integrated development environment (IDE) on steroids. Developed in Rust by Jacob Jackson when he was a student at the University of Waterloo, Tabnine has evolved into a fully-fledged, AI-based code completion tool. Tabnine supports over 20 languages and 15 editors, including popular IDEs like VS Code, IntelliJ, Android Studio, and even Vim. It is available at the price of $432 per year for a team of 3 developers.

2.5 CodeT5

CodeT5 is an open source programming language model built by researchers at SalesForce. It is based on Google's T5 (Text-to-Text Transfer Transformer) framework. In order to train CodeT5, the team sourced over 8.35 million instances of code, including user comments, from publicly accessible GitHub repositories. A majority of these datasets were derived from the Code Search Net dataset, which includes Ruby, JavaScript, Go, Python, PHP, C, and C#, in addition to two C and C# datasets from Big Query.

CodeT5 can potentially bring three capabilities to software programming:

Text-to-code generation: generate code based on the natural language description

Code auto completion: complete the whole function of code given the target function name

Code summarization: generate the summary of a function in natural language description

2.6 Polycoder

Polycoder is an open source alternative to OpenAI's Codex. Developed by the researchers at Carnegie Mellon University, the model is based on OpenAI's GPT-2, which is trained on a 249 GB codebase written in 12 programming languages. According to PolyCoder's authors, the program is capable of writing C with greater accuracy than any other model, including Codex.

While most of the code generators are not open source, Polycoder is one of the first open source code generation models.

2.7 Cogram

Cogram, a Y-Combinator, Berlin-based Startup, is a code generation tool aimed at data scientists and Python programmers using SQL queries and Jupyter Notebooks. Data scientists can write queries in the English language that the tool translates into complex SQL queries with joins and grouping. It supports SQLite, PostgreSQL, MySQL, and Amazon Redshift. Python and Julia developers can integrate Cogram with Jupyter Notebooks to auto-generate code. The tool can generate contextual code for a specific task based on the comments. Data scientists can even generate visualizations based on mainstream Python modules such as Matplotlib, Plotly, or Seaborn.

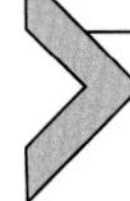

3. Materials and methods

3.1 Deep neural networks and tokenization models used

Regarding the DNN architectures employed, we chose to use the following ones: Average Stochastic Gradient Descent (ASGD) Weight-Dropped LSTM (AWD-LSTM) [1], QuasiRecurrent Neural Networks (QRNNs) [2], and Transformer [3]. These DNN architectures have reportedly obtained some state-of-the-art results [1,4–16] recently in the NLP field in several ground breaking digital products and in some of the most known datasets like the Penn Tree Bank [17], WikiText-2 and WikiText-103 [18], the One-Billion Word benchmark [19], or The Hutter Prize Wikipedia dataset.

The AWD-LSTM is a variation of the famous Long Short-Term Memory (LSTM) architecture [20]. The LSTM is a type of Recurrent Neural Network (RNN) especially capable of processing and predicting sequences. That ability with sequences is the reason why LSTMs have been employed largely in LMs [21]. The AWD-LSTM includes several optimizations compared to the regular LSTM. Two of the most important ones are the use of Average Stochastic Gradient Descent (ASGD) and the weight dropout. The ASGD is used as the NN's optimizer to consider the previous weights (not only the current one) during each training iteration.

The weight dropout introduces the dropout technique [22] to avoid overfitting, but with the characteristic of returning zero, not with a subset of activations between layers, like in traditional dropout, but with a random subset of weights.

The QRNN is another type of RNN that includes alternate convolutional and pooling layers in the architecture. This design makes the QRNN able to capture better long-term sequences and train much faster since the convolutional layers allow the computation of intermediate representations from the input in parallel. They can be up to 16 times faster at training and test time than LSTMs while having better predictive accuracy than stacked LSTMs of the same hidden size. We use a modified QRNN (AWD-QRNN) to include the same ASGD and weight dropout modifications to improve its stability and optimize its capabilities, as for the AWD-LSTM. We utilize AWD-LSTM and AWD-QRNN to produce LMs capable of solving the task of generating source code based on the input as in the literature [1,4–8,12–14].

Transformer is probably the most popular current DNN architecture in NLP due to its performance and recent state-of-the-art results in many tasks. It is an encoder-decoder architecture in which each layer uses attention mechanisms. This use of (self-)attention mechanisms makes Transformer able to model the relationships between all words in a sentence regardless of their respective position. That implies a significant improvement over RNNs, enabling much more parallelization of data processing and unblocking the training over more massive datasets. The excellent results of the Transformer architecture empowered the NLP community to create new state-of-the-art transformer-based models [23] like those used in the current research: GPT-2 [24], BERT [16], and RoBERTa [25].

We chose to use GPT-2 since it is a causal transformer (unidirectional) that can predict the next token in a sequence. Therefore, it can generate source code based on the input, allowing us to compare the results with the AWD-LSTM and AWD-QRNN experiments. Regarding BERT and RoBERTa, we used them to study how a masked modeling approach can auto-complete the source code. In that case, we did not use them for text generation, as in the other experiments, since BERT and RoBERTa are not designed for text generation.

However, they can generate text (more diverse, but slightly worse in quality) [26]. Considering the tokenization techniques, for every AWD-LSTM and AWD-QRNN, we chose the following types of tokens: word, unigram, char, and Byte-Pair Encoding (BPE) [27]—albeit, some studies showed that BPE is suboptimal for pre-training [28]. For the Transformer models, we used the default ones from the pre-defined models: the word piece method [29] for BERT and BPE over raw bytes instead of Unicode characters for GPT-2 and RoBERTa. The different techniques were selected because they produce different token granularities that can enrich our experimentation: full words, sub-words of specific sizes, character-sized tokens, or byte pairs. Furthermore, they enable us to compare the tokenization between the different types of models and tasks to solve.

4. Experimentation details

The dataset used for the experimentation is the Python dataset included in the "GitHub CodeSearchNet Challenge dataset" [30]. It includes 2 million (comment, Python code) pairs from open-source libraries. As observed during the dataset preparation for our experiments, there are about 11 million Python code sentences. The reason for choosing this dataset is that it has already been used in previous research related to NLP and source code. The full dataset includes several languages: Go, Java, JavaScript, PHP, Python, and Ruby. We chose to use only the Python part of the dataset because it enables us to compare the existing literature, which uses the Python language more than other programming languages. The software libraries and packages used primarily during the research were the following: FastAI [31], Google SentencePiece [32], and Hugging Face's Transformers [33].

The preprocessing applied to the dataset included removing most of the code comments and auto-formatting the code according to the PEP-8 Python style guide using the autopep8 package. Regarding the AWD-LSTM networks, we used the FastAI-provided base models pre-trained using the Wikitext-103 dataset [18]. There are no default pre-trained models in FastAI's AWD-QRNN version of those networks, so we trained them from scratch. Regarding the Transformer architectures, we used three standard pre-trained models as a basis: GPT-2, BERT, and RoBERTa. In each case, the exact pre-trained model used were gpt2, bert-base-cased, and roberta-base. These pre-trained models are available from Hugging Face's model. As the reader could infer from the previous explanations about using pre-trained versions, we followed a transfer learning approach similar to other researchers in existing literature [13,14,34,35].

We employed the pre-trained models on English texts to later fine-tune the models for the selected tasks using the GitHub CodeSearchNet dataset. The deep neural network-related source code was coded using the FastAI library (Versions 1.0.61 and 2 dev 0.0.21). To apply the different tokenization techniques to the AWD-LSTMs and QRNNs, we replaced the default Spacy tokenizer [36] with Google SentencePiece [32], following a similar approach to [37]. In the development of the Transformer architectures to see how they perform in filling in the blanks and generating texts, we used Hugging Face's Transformer library combined with FastAI v2 as included on the code repository that supports this paper. To train the neural networks, we used some techniques that are worth mentioning (all the details are in the code repository). To find the most appropriate learning rate to use automatically, we used the function lr_find provided by FastAI following the proposal of [38]. This function trains the DNN over the dataset for a few iterations while varying the learning rates from very low to very high at the beginning of each mini-batch of data to find which is the optimal one regarding the error (loss) metrics until the DNN diverges.

To pursue a faster convergence, we scheduled the learning rate as described in [39] using the one cycle policy (fit_one_cycle) in FastAI. Considering the transfer learning technique used, we trained the first "one cycle" on the top of the existing pre-trained model to later unfreeze all the model layers and perform a more extended training (10–30 epochs) to improve the results. Regarding other training details, in general, we used the default parameters from FastAI, except for a fixed multiplier to control all the dropouts (drop_mult) in the AWD-LSTMs and AWD-QRNNs set to 0.3 because of some heuristics discovered during testing. Furthermore, we decided to train similar architectures using a fixed number of epochs to make the models comparable. For the AWD-LSTM and AWD-QRNN, we used 30 epochs for fine-tuning because we found during the experimentation that the most remarkable improvement for every model produced occurs during that range of iterations.

Similarly, we fine-tuned the transformers for ten epochs since we did not find a significant improvement after that. For more information about the training setup and software details, please refer to the repository that supports this paper and the FastAI documentation. Finally, the hardware used to run the different software and neural network training was a computer running Ubuntu Linux 18.04 LTS Bionic Beaver (64 bits). It has two Nvidia Tesla V100 GPU x16 gigabytes of RAM (Nvidia CUDA Version 10.1), a CPU with 16 cores Intel(R) Xeon(R) CPU E5-2690 v4 @ 2.60GHz, 120 gigabytes of RAM, and 120 gigabytes for the primary disk (HDD) All the supporting materials and software details related to this paper are publicly available in a GitHub repository [40]. The NN models produced are under the Zenodo record [41].

5. Results

This section presents the results achieved after the full training of the selected DNN architectures with the different tokenization models. As outlined in the previous section, we trained the AWD-LSTM and AWD-QRNN DNN architectures using different tokenization model—word, unigram, BPE, and char—and Transformer using three different base models (GPT-2, BERT, and RoBERTa). We trained every AWD-LSTM and AWD-QRNN using one epoch to fit the model's head and fine-tuned for 30 epochs. Meanwhile, the Transformer networks were trained equally for one epoch to fit the head and fine-tune the models for ten epochs. We followed a two-way strategy to evaluate the NNs trained: using the NN training metrics and human evaluation of the models' output.

The metrics used are some of the most common in the literature: accuracy for the validation set and loss for the training and validation sets. They help researchers understand how the NN acts over time, how the model is fitted to the dataset, and the performance and error scores while using training and validation datasets. In this case, the accuracy is the score concerning the LM's ability to predict the next word of filling in the missing ones accurately given a sequence of words from the validation set. The loss metrics reports the error after applying the DNN to the training or validation set, respectively. Every implementation detail related to the DNNs and the metrics is available in the GitHub repository [40]. Apart from those metrics, we assessed the models' quality by applying them in the proposed tasks—generate text and auto-complete—and observing how they performed.

6. Training results

Table 1 displays the final metrics for the different NNs at the end of the training. Similarly, Figs. 1 and 2 show the evolution of each model's accuracy during the training. Figures 3–6 show the training loss and validation loss evolution along the training epochs.

Table 1 Results after full training of each NN architecture.

DNN architecture	Epochs	Accuracy	Train loss	Validation loss	Pre-trained?
AWD-LSTM word	31	0.494893	2.308937	2.341698	yes
AWD-LSTM unigram	31	0.557226	1.639875	1.826841	yes
AWD-LSTM BPE	31	0.580373	1.561393	1.703536	yes
AWD-LSTM char	31	0.779633	0.757956	0.742808	yes
AWD-QRNN word	31	0.515747	1.972508	2.144126	No
AWD-QRNN unigram	31	0.539951	1.790150	1.894901	No
AWD-QRNN BPE	31	0.538290	1.824709	1.896698	No
AWD-QRNN char	31	0.736358	0.944526	0.897850	No
GPT-2	11	0.743738	1.407818	1.268246	yes
BERT	11	0.999238	0.014755	0.004155	yes
RoBERTa	11	0.999468	0.010569	0.002920	yes

AWD, Average Stochastic Gradient Descent Weight-Dropped; BPE, Byte-Pair Encoding; QRNN, Quasi-Recurrent Neural Network

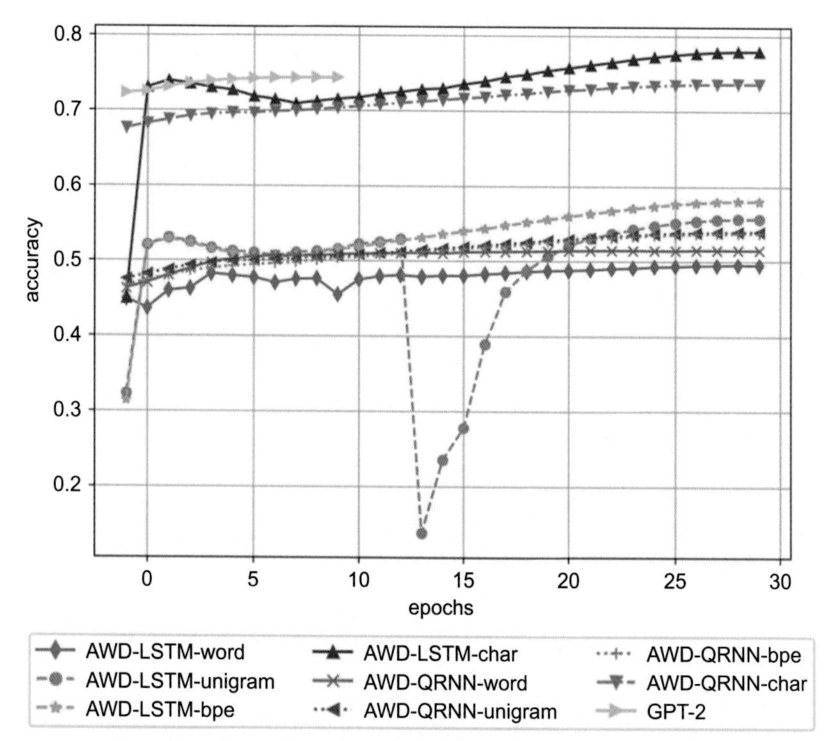

Fig. 1 Evolution of the accuracy of neural networks devoted to source code generation during the training epochs.

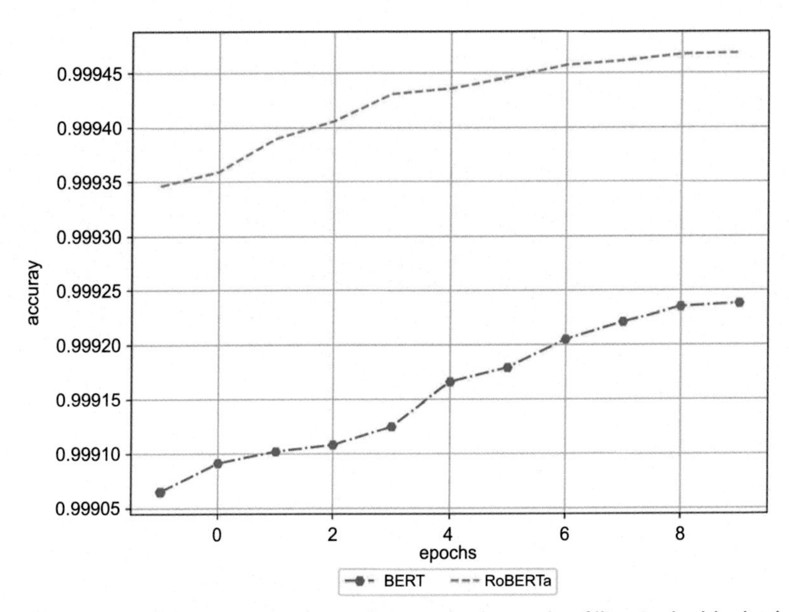

Fig. 2 Evolution of the accuracy of neural networks devoted to filling in the blanks during the training epochs.

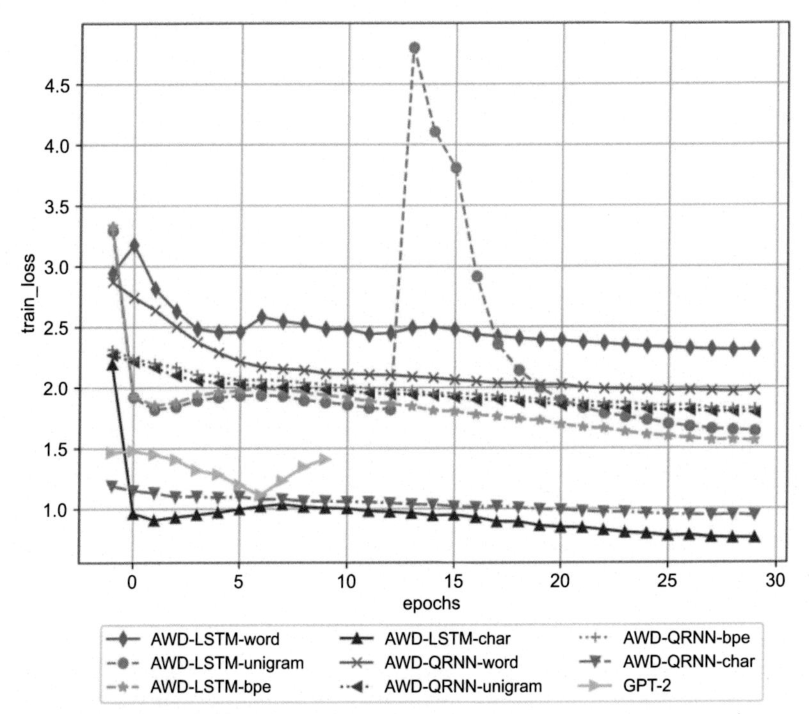

Fig. 3 Evolution of the training loss of DNNs devoted to generating source code during the training epochs.

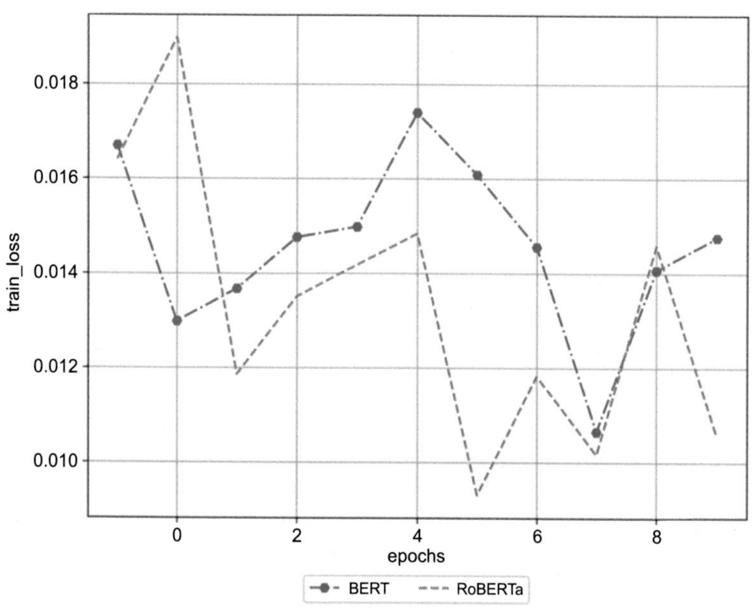

Fig. 4 Evolution of the training loss of neural networks devoted to filling in the blanks during the training epochs.

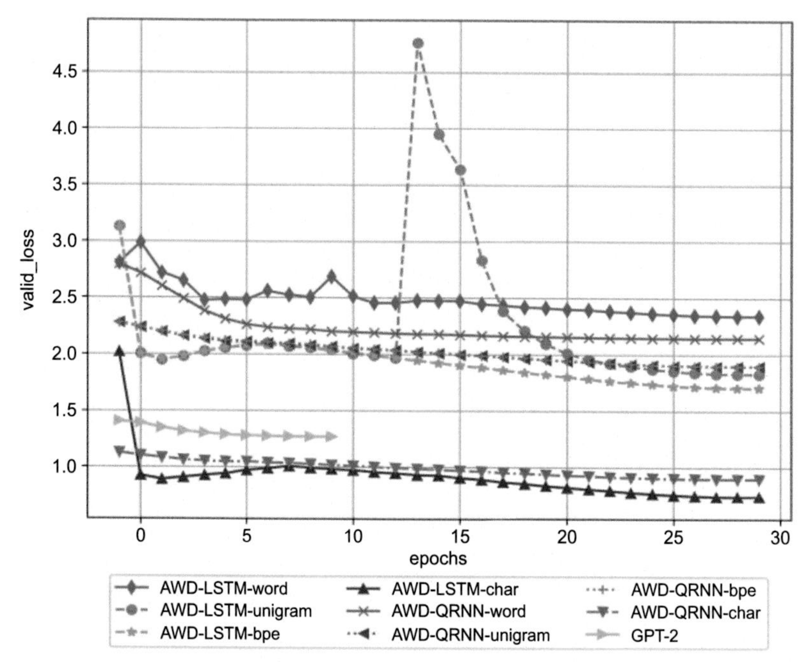

Fig. 5 Evolution of the validation loss of DNNs devoted to generating source code during the training epochs.

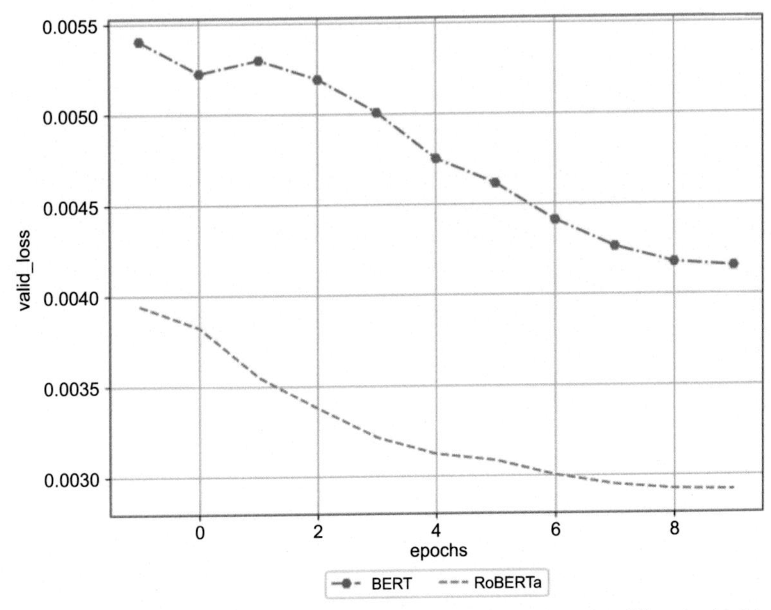

Fig. 6 Evolution of the validation loss of neural networks devoted to filling in the blanks during the training epochs.

On the one hand, according to the results displayed in Table 1 and Fig. 1, for neural networks intended for automated source code generation—AWD-LSTM, AWDQRNN, and Transformer GPT-2—the overall NN-tokenization model combination that performed better in the case of accuracy metrics was the AWD-LSTM with char tokenization (accuracy 0.779633). The second one was the GPT-2 transformer model—BPE over raw bytes tokenization—(0.743738) and the next AWD-QRNN with char tokenization (0.736358). Related to the AWD-LSTM and AWD-QRNN architectures' combination with other tokenization techniques, we obtained poor results on accuracy: between 0.494893 and 0.580373. On the other hand, according to the results shown in Table 1 and Fig. 2, both models (BERT and RoBERTa) had excellent accuracy results in the Transformer models intended for auto-completion, 0.999238 and 0.999468, respectively. Regarding how the pre-training and transfer learning affected the results, the two top results regarding the accuracy were the pre-trained models in the English language (0.779633 and 0.743738), yet the third best result was from a non-pre-trained network (0.736358). Comparing the similar networks, the average (mean) accuracy of the AWDLSTM pre-trained versions was 0.603031 (standard deviation (std) of 0.123144), while the average accuracy of the AWD-QRNN non-pre-trained versions was 0.582587 (std of 0.103107). The only combination NN-tokenization model that worked worse when it was pre-trained was the one with the word tokenization.

Regarding the observed losses, it is worth commenting that the AWD-LSTM char, AWD-QRNN char, and the three transformer models (GPT-2, BERT, RoBERTa) could be trained for more epochs or with a higher learning rate. The model may have been underfitting since the training loss was higher than the validation loss (Table 1, Figs. 3–6). To put the accuracy achieved during the experimentation into context, we compare the results with the existing literature. The papers [42–44] presented models trained to generate Python code that were considered the state-of-the-art when they were published. Our three best models trained to generate source code outperformed the approaches based on the following architectures: vanilla LSTM (accuracy: 67.3%) [43], attention LSTM (69.8%) [43], pointer mixture network (70.1%) [43] or probabilistic model (69.2%) [44].

Our models performed worse than the other approaches based on the Abstract Syntax Tree (AST) instead of using the textual code: AST-LSTM (accuracy 90.3%) [42] or ASTMLP (90.1%) [42]. Therefore, considering this, our results are good in terms of accuracy compared to the existing peer-reviewed literature. As a side note, we did not find models available publicly to compare with, neither in these, nor in other peer-reviewed papers. For the models devoted to auto-complete code (BERT and RoBERTa), as of now, we did not find peer-reviewed papers trying to solve this task using Transformer architectures.

7. Discussion

Considering the results obtained, one could convincingly assert that the tokenization model used profoundly affects the results when generating automated source code.

7.1 Discussing the outcomes from the resulting models

First, our overall results are consistent with the existing literature [21,45–48]. Sub-word tokenization works better in the case of modeling source code, as [45,46] stated. Every result obtained is consistent in that sense. Even more, as [48] envisioned, char tokenization probably should the best option to try by default when dealing with LMs and source code. Furthermore, according to the results achieved, models such as GPT-2—using a tokenization model based on BPE over raw bytes—can outperform LSTM/QRNN models like those we tested to grasp better the internal aspects of a programming language. As showcased during the results, even if GPT-2 didn't best suit in terms of accuracy, it gave better code outputs than the other ones selected for the comparison. As future work, it would be great to check if the better textual output in the case of GPT-2 is because of (a) it being a much bigger and better pre-trained model (163,037,184 parameters against 36,491 for AWD-LSTM and AWD-QRNN models), (b) it being related to the dataset's size or quality, (c) if this is related to both causes, or (d) if it this related to other issues. Continuing with the comments about the accuracy, one may note that the textual outputs generated by the AWD-LSTM char, AWD-QRNN char, and GPT-2 could be polished to be more accurate. The final training loss is higher than the validation loss for the three selected DNN architectures, which can be a sign of under fitting. We found the same issue (train loss « valid_loss) for the BERT- and RoBERTa-based models. While the purpose of this paper is not to produce state-of-the-art results per se, we continued the training for over five more epochs to verify it. The improvement obtained from extending the training for the best approaches was residual in general, so we decided to report the results for 1 + 30 epochs for the AWD-LSTMs and AWD-QRNNs and 1 + 10 epochs for Transformer.

8. Conclusions

This chapter compares how different approaches to tokenization models, deep neural network architectures, pre-trained models, and transfer learning affect the results from language models used to generate source code or auto-complete software pieces. We studied different DNN architectures like AWD-LSTM, AWD-QRNN, and Transformer to seek which kind works better with different tokenization models (word, unigram, BPE, and char). Furthermore, we compared the pre-training effect on the results given by LMs after training them and fine-tuning them via transfer learning to work with other languages (English language to Python programming language). As a result of this work, we find that in small LMs (like our AWD-LSTM and AWD-QRNN models), the tokenization using char-sized chunks works better than using any other tokenization models. In larger models like Transformer GPT-2, the accuracy was slightly worse than the other architectures. However, GPT-2 gave better results on the source code generation tests (even using another tokenization approach like

BPE over raw bytes). For source code auto-completion, we tested some Transformer models like BERT and RoBERTa. While their accuracy was above any other models, they did not perform very well when performing the tasks proposed in our tests. In general, we find that pre-trained models work better, though they were not trained initially for a programming language like Python (our models were pre-trained using the English language). Finally, related to evaluating tasks like automating source code generation and source code auto-completion, we raise concerns about the literature gaps and propose some research lines to work on in the future.

References

[1] C. Gong, D. He, X. Tan, T. Qin, L. Wang, T.Y. Liu, Frage: frequency-agnostic word representation, Adv. Neural Inf. Process. Syst. 31 (2018) 1334–1345.

[2] S. Hochreiter, J. Schmidhuber, Long short-term memory, Neural Comput. 9 (1997) 1735–1780.

[3] M. Honnibal, I. Montani, S. Van Landeghem, A. Boyd, spaCy: Industrial-strength natural language processing in Python, Zenodo (2020).

[4] J. Howard, S. Gugger, Fastai: a layered API for deep learning, Information 11 (2020) 108.

[5] J. Howard, S. Ruder, Universal language model fine-tuning for text classification, in: Proceedings of the 56th Annual Meeting of the Association for Computational Linguistics (Volume 1: Long Papers), Melbourne, Australia, 15–20 July, 2018, pp. 328–339.

[6] H. Husain, H.H. Wu, T. Gazit, M. Allamanis, M. Brockschmidt, Codesearchnet challenge: Evaluating the state of semantic code search, arXiv (2019). arXiv:1909.09436.

[7] International Conference on Learning Representations, Vancouver, BC, Canada, 30 April–3 May 2018.

[8] R.M. Karampatsis, H. Babii, R. Robbes, C. Sutton, A. Janes, Big Code != Big vocabulary: open-vocabulary models for source code, in: Proceedings of the ACM/IEEE 42nd International Conference on Software Engineering, New York, NY, USA, 24 June–16 July, 2020, pp. 1073–1085.

[9] R.M. Karampatsis, C. Sutton, Maybe deep neural networks are the best choice for modeling source code, arXiv (2019). arXiv:1903.05734.

[10] A. Karpathy, The unreasonable effectiveness of recurrent neural networks, Andrej Karpathy Blog 21 (2016) 23.

[11] Y. Kim, Y. Jernite, D. Sontag, A.M. Rush, Character-aware neural language models, in: Proceedings of the AAAI'16: Thirtieth AAAI Conference on Artificial Intelligence, Phoenix, AZ, USA, 12–17 February 2016; AAAI Press: Menlo Park, CA, USA, 2016, pp. 2741–2749.

[12] B. Krause, E. Kahembwe, I. Murray, S. Renals, Dynamic evaluation of neural sequence models, in: Proceedings of the 35th International Conference on Machine Learning, Stockholm, Sweden, 10–15 July 2018; Volume 80, 2018, pp. 2766–2775.

[13] T. Kudo, J. Richardson, SentencePiece: a simple and language independent subword tokenizer and detokenizer for Neural Text Processing, in: Proceedings of the 2018 Conference on Empirical Methods in Natural Language Processing: System Demonstrations, Brussels, Belgium, 31 October–4 November, 2018, pp. 66–71.

[14] J. Li, Y. Wang, M.R. Lyu, I. King, Code completion with neural attention and pointer networks, in: Proceedings of the Twenty-Seventh International Joint Conference on Artificial Intelligence (IJCAI-18), Melbourne, Australia, 19–25 August, 2017, pp. 4159–4165.

[15] Y. Liu, M. Ott, N. Goyal, J. Du, M. Joshi, D. Chen, O. Levy, M. Lewis, L. Zettlemoyer, V. Stoyanov, Roberta: A robustly optimized bert pretraining approach, arXiv (2019). arXiv:1907.11692.
[16] S. Merity, N.S. Keskar, R. Socher, An analysis of neural language modeling at multiple scales, arXiv (2018). arXiv:1803.08240.
[17] S. Merity, N.S. Keskar, R. Socher, Regularizing and optimizing LSTM language models, in: Proceedings of the International Conference on Learning Representations, 2018, Vancouver, BC, Canada, 30 April–3 May, 2018.
[18] S. Merity, C. Xiong, J. Bradbury, R. Socher, Pointer sentinel mixture models, in: Proceedings of the 5th International Conference on Learning Representations (ICLR 2017), Toulon, France, 24–26 April, 2017.
[19] T. Mikolov, A. Deoras, S. Kombrink, L. Burget, J. Cernocky, Empirical evaluation and combination of advanced language modeling techniques, in: Proceedings of the Twelfth Annual Conference of the International Speech Communication Association, Florence, Italy, 27–31 August, 2011.
[20] A.T. Nguyen, T.N. Nguyen, Graph-based statistical language model for code, in: Proceedings of the 2015 IEEE/ACM 37th IEEE international conference on software engineering, Florence, Italy, 24 May, 2015, pp. 858–868. Volume 1.
[21] Y. Ganin, E. Ustinova, H. Ajakan, P. Germain, H. Larochelle, F. Laviolette, M. Marchand, V. Lempitsky, Domain-adversarial training of neural networks, J. Mach. Learn. Res. 17 (2016) 1–35.
[22] A.T. Nguyen, T.T. Nguyen, T.N. Nguyen, Lexical statistical machine translation for language migration, in: Proceedings of the 2013 9th Joint Meeting on Foundations of Software Engineering, Saint Petersburg, Russia, 18 August, 2013, pp. 651–654.
[23] Y. Oda, H. Fudaba, G. Neubig, H. Hata, S. Sakti, T. Toda, S. Nakamura, Learning to generate pseudo-code from source code using statistical machine translation (t), in: Proceedings of the 2015 30th IEEE/ACM International Conference on Automated Software Engineering (ASE), Lincoln, NE, USA, 9–13 November, 2015, pp. 574–584.
[24] K. Papineni, S. Roukos, T. Ward, W.J. Zhu, BLEU: A method for automatic evaluation of machine translation, in: Proceedings of the 40th annual meeting of the Association for Computational Linguistics, Philadelphia, PA, USA, 7–12 July, 2002, pp. 311–318.
[25] M. Post, A call for clarity in reporting BLEU scores, in: Proceedings of the Third Conference on Machine Translation: Research Papers, Belgium, Brussels, 31 October–1 November 2018; Association for Computational Linguistics: Brussels, Belgium, 2018, pp. 186–191.
[26] S. Proksch, J. Lerch, M. Mezini, Intelligent code completion with Bayesian networks, ACM Trans. Softw. Eng. Methodol. (TOSEM) 25 (2015) 1–31.
[27] A. Radford, J. Wu, R. Child, D. Luan, D. Amodei, I. Sutskever, Language models are unsupervised multitask learners, OpenAI Blog 1 (2019) 9.
[28] J.W. Rae, A. Potapenko, S.M. Jayakumar, C. Hillier, T.P. Lillicrap, Compressive transformers for long-range sequence modelling, in: Proceedings of the International Conference on Learning Representations, 2019, New Orleans, LA, USA, 6–9 May, 2019.
[29] V. Raychev, P. Bielik, M. Vechev, Probabilistic model for code with decision trees, ACM SIGPLAN Not. 51 (2016) 731–747.
[30] M.T. Ribeiro, T. Wu, C. Guestrin, S. Singh, Beyond accuracy: behavioral testing of NLP models with checklist, in: Proceedings of the 58th Annual Meeting of the Association for Computational Linguistics, Association for Computational Linguistics, Online, 5–10 July, 2020, pp. 4902–4912.
[31] B. Roziere, M.A. Lachaux, L. Chanussot, G. Lample, Unsupervised translation of programming languages, Adv. Neural Inf. Process. Syst. (2020) 33.

[32] S. Ruder, M.E. Peters, S. Swayamdipta, T. Wolf, Transfer learning in natural language processing, in: Proceedings of the 2019 Conference of the North American Chapter of the Association for Computational Linguistics: Tutorials, Minneapolis, MN, USA, 2 June, 2019, pp. 15–18.
[33] M. Schuster, K. Nakajima, Japanese and korean voice search, in: Proceedings of the 2012 IEEE International Conference on Acoustics, Speech and Signal Processing (ICASSP), Kyoto, Japan, 25–30 March, 2012, pp. 5149–5152.
[34] R. Sennrich, B. Haddow, A. Birch, Neural machine translation of rare words with subword units, in: Proceedings of the 54th Annual Meeting of the Association for Computational Linguistics (Volume 1: Long Papers), Berlin, Germany, 7–12 August, 2016, pp. 1715–1725.
[35] L.N. Smith, Cyclical learning rates for training neural networks, in: Proceedings of the 2017 IEEE Winter Conference on Applications of Computer Vision (WACV), Santa Rosa, CA, USA, 24–31 March, 2017, pp. 464–472.
[36] L.N. Smith, N. Topin, Super-convergence: very fast training of neural networks using large learning rates, in: Proceedings of the Artificial Intelligence and Machine Learning for Multi-Domain Operations Applications. International Society for Optics and Photonics, Baltimore, MD, USA, 15–17 April, 2019. Volume 11006, p. 1100612.
[37] N. Srivastava, G. Hinton, A. Krizhevsky, I. Sutskever, R. Salakhutdinov, Dropout: a simple way to prevent neural networks from overfitting, J. Mach. Learn. Res. 15 (2014) 1929–1958.
[38] S. Takase, J. Suzuki, M. Nagata, Direct output connection for a high-rank language model, in: Proceedings of the 2018 Conference on Empirical Methods in Natural Language Processing, Brussels, Belgium, 31 October–4 November, 2018, pp. 4599–4609.
[39] R. Tiwang, T. Oladunni, W. Xu, A deep learning model for source code generation, in: Proceedings of the 2019 SoutheastCon, Huntsville, AL, USA, 11–14 April, 2019, pp. 1–7.
[40] A. Vaswani, N. Shazeer, N. Parmar, J. Uszkoreit, L. Jones, A.N. Gomez, Ł. Kaiser, I. Polosukhin, Attention is all you need, in: Advances in Neural Information Processing Systems, The MIT Press, Cambridge, MA, USA, 2017, pp. 5998–6008.
[41] A. Wang, K. Cho, BERT has a mouth, and it must speak: BERT as a Markov random field language model, in: Proceedings of the Workshop on Methods for Optimizing and Evaluating Neural Language Generation (NAACL HLT 2019), 2019, Minneapolis, MN, USA, 6 June, 2019, pp. 30–36.
[42] C. Chelba, T. Mikolov, M. Schuster, Q. Ge, T. Brants, P. Koehn, T. Robinson, One billion word benchmark for measuring progress in statistical language modeling, in: Proceedings of the Fifteenth Annual Conference of the International Speech Communication Association, Singapore, 14–18 September, 2014.
[43] J. Cruz-Benito, S. Vishwakarma, NN Models Produced by Cbjuan/Tokenizers-Neural-Nets-2020-Paper: v1.0, Zenodo, 2020.
[44] D. Wang, C. Gong, Q. Liu, Improving neural language modeling via adversarial training, in: Proceedings of the 36th International Conference on Machine Learning, Long Beach, CA, USA, 9–15 June 2019; Chaudhuri, K., Salakhutdinov, R., Eds.; PMLR: Long Beach, CA, USA, 2019, pp. 6555–6565. Volume 97.
[45] J. Devlin, M.W. Chang, K. Lee, K. Toutanova, BERT: pre-training of deep bidirectional transformers for language understanding, in: Proceedings of the NAACL-HLT (1), Minneapolis, MN, USA, 2–7 June, 2019.
[46] C. Donahue, M. Lee, P. Liang, Enabling language models to fill in the blanks, in: Proceedings of the 58th Annual Meeting of the Association for Computational Linguistics, Association for Computational Linguistics, Online, 5–10 July, 2020, pp. 2492–2501.
[47] J. Eisenschlos, S. Ruder, P. Czapla, M. Kadras, S. Gugger, J. Howard, MultiFiT: efficient multi-lingual language model finetuning, in: Proceedings of the 2019 Conference

on Empirical Methods in Natural Language Processing and the 9th International Joint Conference on Natural Language Processing (EMNLP-IJCNLP), Hong Kong, China, 3–7 November 2019; Association for Computational Linguistics: Hong Kong, China, 2019, pp. 5702–5707.
[48] S. Ghosh, P.O. Kristensson, Neural networks for text correction and completion in keyboard decoding, arXiv (2017). arXiv:1709.06429.

Further reading

[49] M. Allamanis, E.T. Barr, C. Bird, C. Sutton, Suggesting accurate method and class names, in: Proceedings of the 2015 10th Joint Meeting on Foundations of Software Engineering, Bergamo, Italy, 30 August–4 September, 2015, pp. 38–49.
[50] M. Allamanis, E.T. Barr, P. Devanbu, C. Sutton, A survey of machine learning for big code and naturalness, ACM Comput. Surv. (CSUR) 51 (2018) 1–37.
[51] A. Baevski, M. Auli, Adaptive input representations for neural language modeling, in: Proceedings of the International Conference on Learning Representations, Vancouver, BC, Canada, 30 April–3 May 2018, 2018.
[52] P. Bielik, V. Raychev, M. Vechev, PHOG: probabilistic model for code, in: Proceedings of the 33rd International Conference on International Conference on Machine Learning, New York, NY, USA, 19–24 June, 2016, pp. 2933–2942.
[53] K. Bostrom, G. Durrett, Byte pair encoding is suboptimal for language model pre-training, in: Proceedings of the Findings of the Association for Computational Linguistics: EMNLP 2020, Association for Computational Linguistics, Online, 16–20 November, 2020, pp. 4617–4624.
[54] J. Bradbury, S. Merity, C. Xiong, R. Socher, Quasi-recurrent neural networks, in: Proceedings of the 5th International Conference on Learning Representations (ICLR 2017), Toulon, France, 24–26 April, 2017.
[55] T.B. Brown, B. Mann, N. Ryder, M. Subbiah, J. Kaplan, P. Dhariwal, A. Neelakantan, P. Shyam, G. Sastry, A. Askell, et al., Language models are few-shot learners, Adv. Neural Inf. Process. Syst. (2020) 33.
[56] C. Bryant, T. Briscoe, Language model based grammatical error correction without annotated training data, in: Proceedings of the Thirteenth Workshop on Innovative Use of NLP for Building Educational Applications, New Orleans, LA, USA, 5 June, 2018, pp. 247–253.
[57] A. Celikyilmaz, E. Clark, J. Gao, Evaluation of Text Generation: A Survey, arXiv (2020). arXiv:2006.14799.
[58] H. Chen, T.H.M. Le, M.A. Babar, Deep learning for source code modeling and generation: models, Applications and Challenges, ACM Comput. Surv. (CSUR) 53 (2020) 1–38.
[59] A. Chronopoulou, C. Baziotis, A. Potamianos, An embarrassingly simple approach for transfer learning from pretrained language models, in: Proceedings of the 2019 Conference of the North American Chapter of the Association for Computational Linguistics: Human Language Technologies, Volume 1 (Long and Short Papers), Minneapolis, MN, USA, 2–7 June, 2019, pp. 2089–2095.
[60] J. Cruz-Benito, I. Faro, F. Martín-Fernández, R. Therón, F.J. García-Peñalvo, A deep-learning-based proposal to aid users in quantum computing programming, in: International Conference on Learning and Collaboration Technologies, Springer, Berlin, Germany, 2018, pp. 421–430.
[61] J. Cruz-Benito, S. Vishwakarma, cbjuan/tokenizers-neural-nets-2020- paper: v1.0, Zenodo, 2020.
[62] P. Czapla, J. Howard, M. Kardas, Universal language model fine-tuning with subword tokenization for polish, arXiv (2018). arXiv:1810.10222.

[63] Z. Dai, Z. Yang, Y. Yang, J.G. Carbonell, Q. Le, R. Salakhutdinov, Transformer-XL: attentive language models beyond a fixed-length context, in: Proceedings of the 57th Annual Meeting of the Association for Computational Linguistics, Florence, Italy, 28 July–2 August, 2019, pp. 2978–2988.
[64] J. Weston, A. Bordes, S. Chopra, A.M. Rush, B. van Merriënboer, A. Joulin, T. Mikolov, Towards ai-complete question answering: A set of prerequisite toy tasks, arXiv (2015). arXiv:1502.05698.
[65] T. Wolf, L. Debut, V. Sanh, J. Chaumond, C. Delangue, A. Moi, P. Cistac, T. Rault, R. Louf, M. Funtowicz, et al., Hugging face's transformers: state-of-the-art natural language processing, in: Proceedings of the 2020 Conference on Empirical Methods in Natural Language Processing: System Demonstrations, Association for Computational Linguistics, Online, 16–20 November, 2020, pp. 38–45.
[66] Z. Yang, Z. Dai, R. Salakhutdinov, W.W. Cohen, Breaking the Softmax Bottleneck: a high-rank RNN language model, in: Proceedings of the International Conference on Learning Representations, 2018, Vancouver, BC, Canada, 30 April–3 May, 2018.
[67] T. Young, D. Hazarika, S. Poria, E. Cambria, Recent trends in deep learning based natural language processing, IEEE Comput. Intell. Mag. 13 (2018) 55–75.

About the authors

Preetha Evangeline David is currently working as an Associate Professor and Head of the Department in the Department of Artificial Intelligence and Machine Learning at Chennai Institute of Technology, Chennai, India. She holds a PhD from Anna University, Chennai in the area of Cloud Computing. She has published many research papers and Patents focusing on Artificial Intelligence, Digital Twin Technology, High Performance Computing, Computational Intelligence and Data Structures. She is currently working on Multi-disciplinary areas in collaboration with other technologies to solve socially relevant challenges and provide solutions to human problems.

Anandhakumar is a professor in the Department of Information Technology at Anna University, Chennai. He has completed his doctorate in the year 2006 from Anna University. He has produced 17 PhD's in the field of Image Processing, Cloud Computing, Multimedia Technology and Machine Learning. His ongoing research lies in the field of Digital Twin Technology, Machine Learning and Artificial Intelligence. He has published more than 150 papers indexed in SCI, SCOPUS, WOS, etc.

Printed in the United States
by Baker & Taylor Publisher Services